NATHANAEL-ISRAEL ISRAEL, PhD
Discoverer of the Universe Turbulent Origin Formula
World's #1 Expert at the Demystified Intersection of Science and Faith
Founder of www.Science180Academy.com
www.Israel120.com

BOLDEST SCIENTIFIC FORMULA OF GOD AND CREATION

Live Better by Quickly Mastering the Only Scientific Theory that Evolution, Big Bang, AI, and Some Religions Desperately Ache You Ignore

$$T = \frac{D}{Ve} + \frac{2R\pi}{Vo}$$

OTHER BOOKS BY NATHANAEL-ISRAEL ISRAEL

Turbulent Origin of the Universe
There is Only One Scientific, Simple, Safe, Trustworthy, Unexpensive, Brave, Practical, Nonconformist, Universal, Verifiable Formula that Accurately Decodes the Universe Formation … But You Are Not Using It

Reconciling Science and Creation Accurately
What Science Accurately Teaches about Creation and God's Existence that Atheists, Freethinkers, and even Most Christians Ignore … And How to Demonstrate it Without Taking Sides Between Rationality and Faith

Turbulent Origin of Chemical Particles
Why You Don't Have to Embrace Evolution, Big Bang, or Deny God to Scientifically Prove the Formation of All Chemical Particles

From Science to Bible's Conclusions
How Decoding the Universe-Origin by Properly Revisiting Scientific Data—That Top Scientists Collected but Wrongly Analyzed—Bizarrely led to the 3500 Years Old Biblical Account of Creation

Turbulent Origin of Life
Why You Don't Have to Embrace Evolutionism or Check Your Brain at the Door in the Name of Faith or Science to Accurately Decrypt the Origin of Life Using the Historic Formula of the Universe Formation

How God Created Baby Universe
What Children Must Scientifically Learn Early about the Universe Formation to Avoid Dangerously Abandoning God Later in Life Just Like Most College Students Who Embrace Evolution and Big Bang That Deny Biblical Creation

How Baby Universe Was Born
How to Scientifically Talk to Children about the Universe Formation and They will Know Forever How to Correctly Test the Intersection of Science and Faith

Origin of the Spiritual World
Top Secrets about the Origin of Everything in the Universe that Some Elites Have Hidden from You for Thousands of Years

Science180 Accurate Scientific Proof of God
Can We Scientifically Explain the Formation of the Universe Through Natural Processes Without Evoking Evolution and Big Bang?

Mathematical Proof of God's Existence at the Intersection of Science and Faith'
The Scientifically Verifiable Cosmological Theory that Challenges the Big Bang Theory at the Crossroads of Reason and Religion THEY Want You to Ignore

NATHANAEL-ISRAEL ISRAEL, PhD
Discoverer of the Universe Turbulent Origin Formula
World's #1 Expert at the Demystified Intersection of Science and Faith
Founder of www.Science180Academy.com
www.Israel120.com

BOLDEST SCIENTIFIC FORMULA OF GOD AND CREATION

Live Better by Quickly Mastering the Only Scientific Theory that Evolution, Big Bang, AI, and Some Religions Desperately Ache You Ignore

$$T = \frac{D}{Ve} + \frac{2R\pi}{Vo}$$

Science180
Augusta
United States of America
www.Science180Publishing.com

Visit the author's website at Israel120.com

BOLDEST SCIENTIFIC FORMULA OF GOD AND CREATION
Live Better by Quickly Mastering the Only Scientific Theory that Evolution, Big Bang, AI, and some Religions Desperately Ache YOU Ignore

First edition: April 24, 2026

Published by Science180
Augusta (USA)
www.Science180Publishing.com

Book Cover and Illustrations by Nathanael-Israel Israel
ISBN: 9781969961014 (Paperback)

Library of Congress Control Number: 2026909462

More books by the same author can be found at Israel120.com/books and Science180.com/books

For information about special discounts available for bulk purchases, please visit Science180.com/discount

Science180 can bring authors including Dr. Nathanael-Israel Israel to your live or recorded events. To book him for an event, please visit Science180.com/speaking

To publish your book(s) with Science180, go to Science180Publishing.com

To interview the author of this book, visit Israel120.com/interview
To donate, please visit Israel120.com/donate or Science180.com/donate.

For any questions, please visit Science180.com/contact

Printed in the United States of America.

CONTENT

Section 1: Why and How this Book is Written *1*

1. The one thing I want to achieve with this book and what you can get out of it 2

2. How I developed the scientific method I used to decode the origin of the universe and test the existence of God 4

Section 2: The 4 Simple Scientific Evidences Required to Crack the Origin of the Universe and Test God's Existence Scientifically Without Taking Sides Between Science and Faith *10*

3. Discover the 4 most essential variables and the 3 celestial bodies you must master if you want to confidently decrypt the formation of the universe scientifically 11

4. Ignore the orbital speed of celestial bodies, and you will never break the code of the universe 12

5. The radius of celestial bodies: a master variable to respect if you are serious about finding the origin of the universe 13

6. If you think that the distance separating celestial bodies is by chance, think twice about what is called the semi-major axis 14

7. Understand this simple variable (escape velocity) discovered hundreds of years ago, and you will prove the origin of any celestial body in the universe easily 15

8. He who can pay attention to these 12 simple facts collected over centuries and confirmed by NASA will crack the code of the universe properly. He who cannot will keep wasting time and money cosmologically 16

Section 3: Developing my Turbulent Split-Gathering Methodological Framework: the Foundation of my Inquiry *20*

9. What am I going to tell you in this critical chapter? 21

10. Do all celestial bodies have a common ancestor or precursor? Pay attention to what I called the "Turbulent Primary Materia," which is completely different from what Charles Darwin and evolutionists thought 21

11. He who can master this simple concept of the "split-gathering of precursor of bodies" will have the key that Albert Einstein and Isaac Newton missed! He who cannot will suffer forever 22

12. Understanding the pregnancy of the precursor of celestial bodies is very easy 23

13. The big secret of dealing with the birth of a primary body and its secondary bodies 25

14. Can you ignore the journey of the precursors of the celestial bodies and still understand the universe's origin? No! Therefore, watch this! 27

15. You cannot crack the origin of the universe if you ignore these 2 simple concepts about the journey of the forerunners of the celestial bodies 29
16. Want to accurately know how to calculate the duration of the formation of the celestial bodies easily? Master the Nathanael-Israel Israel Birthdate 51

Section 4: Want to Have a Total Peace of Mind About the Exact Date of Birth of the Earth, Moon, and Sun Without Taking Sides Between Science and Faith? Together, Let Us Simply Evaluate and Apply My Turbulent Split-Gathering Methodological Framework to the Formation of the Earth, Moon, and Sun! 56
17. What is in this section for me?..57
18. Can this Nathanael-Israel Israel Birthdate formula prove that the Earth was formed within days? 57
19. The birthdate of the Moon that no one has ever learned at Harvard University or any Ivy League institution yet..59
20. Proving that the Sun was formed on the 4th day is really easy if you pay attention to the Nathanael-Israel Israel Birthdate of the Sun 61
21. The makeover of the historic story of the formation of the Solar System that took the world by surprise! 64

Section 5: Evaluating and Verifying my Turbulent Split-Gathering Theory and Troubleshooting Failed Theories.. 83
22. To properly test if any existing framework (scientific or religious) partially or fully applies my Turbulent Split-Gathering Framework, just start like this! It is that simple..84
23. Is it easy to test scientifically whether any religion in the world is correct? Pay attention to the code behind the Nathanael-Israel Israel Birthdate of celestial bodies..86
24. Can we use mathematics to scrutinize the Koran and prove that Islam is scientifically incorrect? Can anybody really dare to say that?......................91
25. Why did all the creationists fail to unravel the formula of the universe creation for so long? 96
26. A sure way for Old Earth creationists to make mistakes and miss the historic moment— And how to avoid it? 98
27. What to do when Young Earth creationists and Intelligent Design proponents cannot really claim victory or vengeance? 100
28. What went wrong with atheists, agnostics, and all other freethinkers? 102
29. He who can understand why evolution and the Big Bang theory missed the universe's origin formula has the key to his destiny. He who cannot journeys toward his own perdition..105
30. A sure way to be ungrateful to modern science! How to avoid it 107

Section 6: Science180 Differentiators — Why is the Turbulent Split-Gathering Framework, the Methodological Framework of Science180 Cosmology Novel, and Why Does it Add Value Compared to Existing Frameworks and Theories? *116*
31. What is in this section for me? 117
32. Easy-to-understand universe-origin and God's existence theory 118
33. State-of-the-art decoding experience of the origin of the universe 118
34. The most accurate, reliable, safest, and best explanation of the universe's origin ever 118
35. Universe-origin theory that helps you fight wasteful programs and theories 119
36. Universe-origin rule breaker 119
37. All the universe-origin solutions you love 120
38. Undeniable reconciliation of science and the Biblical account of creation 120
39. Use Science180 Turbulence 1 to test the intersection of science and faith accurately 121

Section 7: Conclusive Remark About my Argument for the Existence of God *125*
40. The "God's Existence Scientific Formula" that everybody is talking about 126
41. Does God exist? What is coming up next after my argument for the existence of God 127

Useful Resource for You *132*
42. Science180 Overview 133
43. Other great books by Dr. Nathanael-Israel Israel 135
44. Which of Nathanael-Israel Israel's books is best for me according to my background? 149

References *151*
Index *152*
About the Author *161*

Section 1: Why and How this Book is Written

"It is impossible to contemplate the spectacle of the starry universe without wondering how it was formed..." — Henri Poincaré

"The most incomprehensible thing about the world is that it is comprehensible." — Albert Einstein

1. The one thing I want to achieve with this book and what you can get out of it

"The people who are crazy enough to think they can change the world are the ones who do." — Steve Jobs

Congratulations, and thank you for your decision to read this book. By the time you finish reading it, you will discover valuable information that will make a real difference in your lives. It will help you perceive the origin of the universe and its content differently, saving you money and time. By freeing your mind from wrong cosmological assumptions, you will improve your health and faith. With a peace of mind, you will confidently answer most of the top-asked questions of all time about the origin of the universe and the scientific proof of God's existence:

- How were the Earth, the Moon, and the Sun really formed?
- Can we scientifically determine the exact date of their formation without making any assumptions?
- Did it really take billions of years to form them, as the Big Bang theory and evolution say, or did God create them as the Bible says?
- Can anyone scientifically demonstrate that the Earth was fully formed by the end of the 3rd day of creation, and the Moon and the Sun on the 4th day, as the Bible says?
- If the Biblical account of creation is true, can anyone prove scientifically that the Biblical days of creation are really 24 hours each rather than billions of years?
- If God created the universe, can math and science collide to demonstrate rationally (in a way that even the smartest atheists or freethinkers can like) which being, among the countless worshipped across the globe, is the true God?
- Are believers (that some freethinkers said are irrational) really proving the existence of God as they say?
- Is science (that some believers say is evil) really against God, whose existence the unbelieving freethinkers say their scientific theories disproved?
- Using a scientifically testable approach, without taking sides between science and faith or engaging in nonsensical theories, can we answer these questions definitively, boldly, rationally, without angering the world's religions, the secular scientific community, the believers, or even God himself (whether you are a believer or not)?

Wow okay! What a daring task! Answering these challenging questions scientifically has been my main assignment for more than 12 years. Because I know you might reasonably expect some get-to-the-point urgency from me, I will not delay my account of discovering the formula for the formation of the universe.

Although everybody wants to know the real answer to these questions, if I answer any of them with a YES or a NO right now, some people will likely stop reading this book because my answer could offend them. Therefore, let mathematics and science guide our rational analysis of existing scientific data on the universe until we arrive at a scientifically verifiable conclusion about who or what really formed or created the universe. Sounds like a good plan? If yes, let's go!

If you survey the answers to these questions across the globe, you can group them into 2 categories:

- What is taught in most public schools (that denies the 6 days of creation by God) and
- What is believed in most religions (that embrace a supreme being in most cases, with creationists accounting for more than 55% of the world's population).

Indeed, for decades, the Big Bang theory has taught that the universe was formed more than 13 billion years ago. In contrast, for about 3500 years, the Bible revealed that the God of Israel finished the formation of the Earth on the 3rd day of creation and the formation of the Moon and the Sun on the 4th day. Other scientific, philosophical, and religious theories and dogmas have their own creation stories. Some religions (e.g., Buddhism) even discourage debates about the formation of the universe. While some Abrahamic believers (Jews, Christians, and Muslims) believe that God created the universe in 6 days. However, it is important to note that there is a significant difference between the Biblical and Islamic creation narratives. Within the creationists, some believe that God used evolutionary processes to create it in billions of years, just as evolutionists and the Big Bang proponents say, while others believe that the days of creation are 24 hours each. Still, some believers support the idea that creation took place over a six-day 24-hour timeline. Without wasting time, how can we scientifically prove who is right? Is there any intersection of science and faith that can accurately tell us how the universe was created without forcing us to take sides between science and faith? We will test that one thing before the end of this book, and you will get the answer scientifically, even on your own.

Well, this is not my first book answering challenging questions about the formation of the cosmos. In fact, I have already devoted 10 books to the origins of the universe, life, and chemical particles, including 9 published in October 2025 and 1 in February 2026 (www.Israel120.com/books). But this book will focus on the very essential facts that nobody can deny. I tested my demonstration over 12 years of intensive research on the matter. The assurance I have in my scientific discovery at the intersection of science and faith (which some perceive as a water-

and-oil relationship) is what led me to label this book the boldest answer to the question of the creation of the universe and the existence of God. In the *Boldest Scientific Formula of God*, I will lead you through the sequence of steps I used to arrive at my conclusion of the origin of the universe. You will learn how I discovered my scientific framework and how I applied it to key celestial bodies and some simple variables to crack the origin of the universe. Without wasting your time, let's get down to it.

2. How I developed the scientific method I used to decode the origin of the universe and test the existence of God

"If you don't like something, change it. If you can't change it, change your attitude." — *Maya Angelou*

Any sound scientific reporting is usually expected to contain most of the following:

- A literature review outlining existing expert perspectives on the topic discussed;
- The presentation of hypotheses regarding specific methods used to address the questions of interest;
- Collection and presentation of the data used; and
- An analysis of the evidence used to prove or disprove the hypotheses.

In this book, I will deal with all those points. Before I submit to your precious attention the bottom line of my findings, I would first like to provide the general context of my research. In fact, I did not initially start this research with a hypothesis, although I could have. But it was my curiosity that led me to NASA's website about 12 years ago to explore what kinds of data were collected about celestial bodies. As I reviewed the data I found there and others I discovered in the literature, I noticed that hundreds of variables are used to characterize celestial bodies. While carefully examining those data, I noticed very interesting trends and patterns, but I could not find any prior work that addressed them. I immediately knew I had a story to tell. To avoid mixing the originality of my story with data produced using puzzling theories, I focused my data collection on raw data directly collected on celestial bodies, and even those hidden in theories I dislike.

After years of investigation, I discovered that they were related to a phenomenon called turbulence, which is considered one of the top unsolved mysteries in science. Turbulence is a state of fluid motion characterized by apparently random, so-called chaotic 3D vortices, or eddies, which are kinds of swirls and twists. In short, to put this in perspective for non-experts, turbulence is related to the motion of highly unstable fluids, such as liquids (e.g., water) and gases (e.g., the air we breathe).

SECTION 1: WHY AND HOW THIS BOOK IS WRITTEN

"Honest" turbulence experts confessed that they really don't know a whole lot for sure about the turbulence phenomenon. Worse, they even disagree a lot about what they think they know. Some turbulence specialists even view most "established" laws and theories of turbulence as more like religious creeds than matters of fact. So, for more than 500 years, experts in turbulence have sought to fully understand it, but little progress has been made. That's why turbulence is classified as one of the most unsolved problems in physics. Hear what William K. George, an expert in turbulence, confessed a few years ago:

"*The lack of a satisfactory understanding of turbulence presents one of the great remaining fundamental challenges to scientists — and to engineers as well, since most technologically important flows are turbulent … When I [Professor William K. George] started working in turbulence in the 1970's, I really thought we would be much farther ahead in our understanding and in our ability to predict them numerically. The fact is that, in my opinion, we have made almost no real progress …. While some experts in the field of turbulence trust some existing theories are a gospel, I [Professor William K. George] am very careful in doing so… Now not everyone who teaches courses on this subject (and especially those who write books about it) will tell you this, but the truth is we really don't know a whole lot for sure about turbulence. And worse, we even disagree about what we think we know! There are indeed some things some researchers think they understand pretty well, but most of them are based on assumptions and logical constructions about how some researchers believe turbulence behaves within certain limits. However, even these ideas have never really been tested in controlled laboratory experiments, because no one has ever had the large-scale facilities required to do so. The acceptance of most ideas in turbulence is perhaps more due to the time elapsed since they were proposed and found to be in reasonable agreement with a limited data base, than that they have been subjected to experimental tests over the range of their assumed validity. Therefore, it might be wise to view most 'established' laws and theories of turbulence as more like religious creeds than matters of fact.*" (George, 2013).

Therefore, when I searched the literature, I did not find any existing methodological framework, meaning a set of ideas or rules, that I could use to study the turbulence I noticed occurring during the formation of the universe, which cosmologists ignore.

Bizarrely, the experts in the field of turbulence don't know that the data they have been searching for in their field and that they are trying to collect on a laboratory scale were already collected on the astronomical scale on celestial bodies with great accuracy. Strangely, the physicists who collected these data to explain the formation of the universe ignore that they already have a treasure that the turbulence experts are looking for, and that borrowing some concepts from the turbulence experts could open their eyes to a new perspective that could crack the code of the universe. In other words, a breakthrough in turbulence was needed, for without it, it would be impossible to understand what the massive data collected on the universe are saying. Because no clear framework exists to explain turbulence, I had to design one myself. In the next chapter, I will deal with just that.

One thing that helped me understand the importance of the data that top scientists collected on celestial bodies throughout the centuries, but that they misinterpreted, was my life story, which led to my interdisciplinary background and to my training to think naturally outside the box more than most people can. If you ignore how I think you may think that I don't know how to think; but my results will speak for me. I don't just follow the crowd and leave my brain at the door in the name of science or faith. I am not afraid to think differently. Because I know I am unique, I will stay unique. It is OK if some people are turned off because of my message. But those (like you) who are meant to hear it will stay. I don't try to be different or sound like others. I know I am different, and what you will read in this book is different from what you were taught or what you think. You may not agree with me yet, but by the time you reach the end of this book, you may shake your head. My friend, I don't, and I will not try to look like or sound like others whom some people are trying to emulate or who are doing similar things like me. Therefore, you will more easily remember my message and me. Those who know me stay on the lookout and search for what I have to offer. Wherever and whenever needed, I will stay contrarian, meaning a person who opposes or rejects popular opinion, or who goes against current modern scientific and religious practices that have kept the truth captive. All this is possible because, among many things, I like to consider the big picture and the long-term impact of my actions. Let me explain to you why part of my education also qualified me to speak like this.

Indeed, although I have a PhD in plant, insect, and microbial science from one of the top universities in the USA, I am also privileged to have a good understanding and expertise in fields beyond life science. This can also justify my full membership in the following professional associations in the USA:

- American Association for the Advancement of Science
- American Chemical Society
- American Society for Microbiology
- American Society of Agricultural and Biological Engineers
- American Society of Agronomy
- American Society of Biochemistry and Molecular Biology
- Crop Science Society of America
- Ecological Society of America
- Soil Science Society of America
- American Physical Society
- American Mathematical Society
- American Philosophical Association

- And many others, including renowned religious, theological, associations / societies. See the full list at Israel120.com/affiliations

I completed my BS and MS in Benin Republic (West Africa)—finishing first in my class—before moving to the USA more than two decades ago, where I also earned a PhD, again finishing first among hundreds of peers (and I served as the Doctoral Marshal at the graduation ceremony). Because of my broader knowledge and interest, it is easier for me to look at things (not just certain scientific data) more deeply than most specialists do, given the narrow understanding some people excuse in their acute specialization. I have a lot to say concerning how my life journey has shaped my unique perspective, but for the sake of space and time, I will defer those details for another opportunity.

Using my mindset to investigate the massive cosmological data collected on the universe, I developed an original method to investigate the turbulence I perceived during the formation of celestial bodies (by the way, these trends are detailed in hundreds of pages in my previous books). With my nonconformist approach, I uncovered overlooked datasets containing answers long sought by both scientific and theological communities. My approach was not based on the Big Bang theory, evolution, or any other existing philosophical theory, including creationist ones. In my books, I detailed how I identified the evidence that informed my methodological framework and how I developed, evaluated, and refined it. In this book, for the sake of time, as I unveil the bottom line of my discovery, I will focus on just 4 essential variables and 3 celestial bodies.

Why just 4 variables and 3 celestial bodies? Because I have demonstrated that their characteristics and the processes connected to them are sufficient to accurately explain the formation of the universe quickly, yet accurately, to busy people (like you) who don't have a lot of time to spare. To avoid confusion, please allow me to first lay some foundations before elaborating on my scientific framework.

Because this book is a scientific report, some people may expect it to adhere to the aforementioned standards. A challenge I faced was how to accommodate both the scientists and the nonscientists in the same book. Because some people who don't understand science are also still looking for scientific proof, I felt like writing a bottom-line book that will strip away all unnecessary complicated details to understand the core message that everybody can get, can be more helpful, knowing that those who are interested in the details can follow up with my other books according to their backgrounds. Additionally, if I have strictly used the standard scientific reporting style here, without keeping nonscientists in mind, some laypeople who are also asking the same questions that highly educated people are requesting and who also deserve an answer would not get much out of it. But before the end of this book, you will clearly see all the elements of scientific reporting.

For instance, I could have begun this book by examining the significance of studying scientific works on the origins of the universe and creation narratives. Still, because I already did that in other books, I will not repeat myself here. In my book "*Turbulent Origin of the Universe*" (Science180.com/scientific), not only did I review what is scientifically known about the origin of the cosmos, but I also demonstrated scientifically everything you will read in this book. In my books *Science180 Accurate Scientific Proof of God* and *Reconciling Science and Creation Accurately*, I also extensively addressed the philosophical and religious accounts of creation worldwide, as well as the variation among creationists. In those books, I broke down the distinctions between belief and ideology, not just between believers and rationalists, but also among the various kinds of creationists, and why believers and their rationalist opponents have failed to convince each other about the formation of the universe or creation.

As I reviewed the literature on the philosophical stories of the formation or origin of the universe, I also discovered that most people (even most rational believers and unbelievers) want scientific proof of God's existence. Still, they don't want to engage in the real math or science that can lead to the sought-after demonstration. While some people requesting scientific proof of God don't even understand science and cannot handle scientific and mathematical demonstrations, others can but don't want to devote the time to reviewing the proof without their own prejudice. Most people are very busy and don't want a long demonstration; they want a simple, rational proof that gets right to the point. Unfortunately, when you give them a short answer, they want the long answer, and when you give them the long answer, they cannot or don't want to read it. So what to do then? Most people have already made up their mind!

One thing that shocked me after the publication of my first 10 groundbreaking books on the origin of the universe was that, noy only do most nonbelievers who claim they are seeking for a rational proof of God's existence already set their mind not to accept any such proof, but also, most believers who say they believe in God don't actually believe that it is possible ever to demonstrate whether God truly created the universe in 6 literal days as a minority of creationists embrace. Most people in that minority think that science is against God. In general, most human beings are not doing the necessary homework to get the answer they are looking for. This attitude is similar to what this Blues song (rendered by Albert King) says:

> *"Everybody wants to go to heaven, but nobody wants to die! Everybody wants to laugh, but nobody wants to cry! Everybody wants to hear the truth, but still they want to tell a lie. Everybody wants to know the reason, but without even asking why …"*

Does that relate to you also? Well, I realized that even if I demonstrated everything here, it wouldn't help most people much. Therefore, although what I discovered is dense, I will simplify it for you, so I can accommodate as many people as I can, knowing that some must endure a necessary pain or sacrifice in the process, so we can transition into or achieve the successful conclusion we are all looking for: use modern science and math accurately to get a definitive answer about the origin of the universe and whether God exist or not, and above all who is He.

As we engage in the scientific journey, please be gracious and patient with me, as I may overexplain certain things (which the most educated may perceive as basic). At the same time, I may overlook others (that the "less" educated may deem too complicated). I have also come to realize that the so-called "less educated" understand many things that the so-called "highly educated" ignore. So, be aware that some details must be omitted to simplify the demonstration. Still, I will do my best to make sure you don't miss the essentials. That is why I call this book the bottom line of my research. Because I am dealing with a very original topic that was less studied before me, I also coined some terms to make it easier for others to replicate my research and align with the vocabulary I used.

Finally, I beg you, please don't say and don't bet that you scientifically know how the origin was formed yet, or that God does not exist, until you test it my way!

In the next chapter, I will explain the raw data I used to reach my conclusion. Then, I will present my methodological framework, which I designed. After that, I will illustrate my discovery with a simple, easy-to-understand makeover of what the scientific data proved really happened during the formation of the universe. Then, using basic math, you will see for yourself if science can really demonstrate the existence of any God or not. If you are ready, let's get down to it.

TAKE-HOME MESSAGE OF SECTION 1: WHY AND HOW THIS BOOK IS WRITTEN

In this chapter, I explained the main questions I want to answer in this book. I also introduced myself and explained how my life story and education prepared me to handle these questions in a unique, never-before-seen way. Considering how people are looking for scientific proof of God or the origin of the universe, yet they don't have the time or inclination to scrutinize the complex facts presented to them, I explained why I chose to write this book without going into too much detail that might overload some readers. Everybody must pay a little price, and all of my detailed demonstrations can be found in my previous books at www.Israel120.com/books.

Section 2: The 4 Simple Scientific Evidences Required to Crack the Origin of the Universe and Test God's Existence Scientifically Without Taking Sides Between Science and Faith

"Number rules the universe." — *Pythagoras*

"Nature is written in mathematical language." — *Galileo Galilei*

3. Discover the 4 most essential variables and the 3 celestial bodies you must master if you want to confidently decrypt the formation of the universe scientifically

"If I have seen further it is by standing on the shoulders of Giants". — Isaac Newton

For centuries, scientists have collected vast amounts of data to understand the origins of the Earth, Moon, and Sun. Many theories have been proposed using some of that data, and more data is being collected, as if the misunderstanding of the origin of the universe were just a matter of insufficient data. What if we already have the data needed to crack the code of the universe, but scientists ignore it? After spending 12 years scrutinizing massive scientific data, I was shocked to discover that, with just 4 variables, one can crack the code of the universe if the right perspective is applied to analyze and interpret them. I also discovered that focusing on just 3 celestial bodies in the universe is sufficient to decode its entire formation. Therefore, focusing on these data, I spearheaded a scientific framework that has finally made the understanding of the formation of the Solar System easier than ever before.

Collected over centuries by some of the top scientists in history, verified and published by NASA and other space agencies across the globe, the data I used in my study were unfortunately misinterpreted by their collectors. Before I explain the essential processes that they imply, I will first list the names of these 4 variables:

- Semi-major axis (i.e., the average distance separating a celestial body from the primary body it orbits)
- Radius (i.e., the straight line from the center to the circumference of a body)
- Orbital speed (i.e., the speed at which a celestial body orbits its primary body)
- Escape velocity (i.e., the minimum speed required for a body to escape the gravity of another body)

The 3 celestial bodies that I will focus on in this book are the Earth, the Moon, and the Sun. Being close to us, these celestial bodies need to be carefully considered when talking about the origin of the universe. But, far from any desire to advance any conspiracy theory here, as you will see in this book, countless variables collected on these bodies were dumped into the scientific garbage as the elites that collected them were hunting for data located at the extremities of the visible universe that we may never have a chance to verify! In my book *Turbulent Origin of the Universe*, I devoted dozens of pages to each of these variables.

Before I elaborate on these variables, I don't want you to think they are the only ones I studied to decode the origin of the universe. In fact, I studied more than 500 variables, some of which helped me understand certain aspects of the gigantic puzzle of the origin of the universe. But if you miss these 4 variables, you will likely never understand the origin of the cosmos, nor be able to test the intersection of science and faith. You may not believe me now, but by the time you finish reading this book, you will be happy for me for breaking down complicated concepts in such simple language that even elementary school students can understand. I hear this all the time not only from people who attend the Science180 Academy (www.Science180Academy.com) that I founded and through which I go deeper with people of all educational background, whether they are creationists (e.g., Science180 creationists, Young Earth creationists, Old Earth creationists, Intelligent Design proponents), or anti-creationists or freethinkers (e.g., atheists, humanists, rationalists, agnostics, non-theists, nonreligious, skeptics, nonbelievers, religiously unaffiliated, spiritual-not-religious, ex-believers, and doubters). Some of the people who fit Science180 Academy have a profound technical knowledge base and a science background, while others don't. Through that program, I deliver outstanding value, insight, and lessons to help people like you accurately understand the true origins of the universe, chemicals, and life, so you can tap into that knowledge to improve lives perpetually. My goal is to give you practical, undeniable proof of the formation of the universe so you can be fired up to become the best version of yourself, and to drive positive changes in your initiatives that will benefit you today and forever. If this is something I can help you with, you may want to check it out. Please allow me to break down each of these 4 variables.

4. Ignore the orbital speed of celestial bodies, and you will never break the code of the universe

"Speed is irrelevant if you are going in the wrong direction." — *Mahatma Gandhi*

The orbital speed of celestial bodies has been known since 1619, when the mathematician Johannes Kepler accurately calculated it in his work on the laws of planetary motion. Kepler proved that, for planets in the Solar System, there is a negative relationship between orbital speed and the distance from the Sun. This means that planets closer to the Sun move faster than those further away. About 70 years after Kepler, in his book called Principia, Sir Isaac Newton proposed a theory of gravity to explain the motion of the planets, thinking that the planets' orbital motion was caused by gravity (by the way, I proved that wrong —see my book *Turbulent Origin of the Universe*). A few centuries later, Albert Einstein came up with a new theory. However, none of these thinkers, nor those who came after them, until my discovery, ever explained the machinery behind these trends of the orbital speed and the distance between the celestial bodies.

By the way, to accommodate the global audience that is interested in this book, I will use the kilometer (abbreviated as km) as the unit of distance rather than the mile, which is mostly used in the USA, where I am based. On average, 1 km is about 0.621 miles, so 1 mile is about 1.61 km. So, when I say km/s, please understand that I mean kilometer per second.

Today, due to advancements in space exploration, the average orbital speed of the 3 main celestial bodies I am focusing on is:

- Earth: 29.78 kilometers /second
- Moon: 1.02 kilometers / second
- Sun: 19.4 kilometers / second

In other words, on average, the Moon turns around the Earth at about 1.02 km/s, the Earth around the Sun at about 29.78 km/s, and the Sun moves about 19.4 km/s relative to nearby stars. For details about why I used the speed of the Sun relative to nearby stars rather than relative to the center of the galaxy, please see my book *Turbulent Origin of the Universe.* But why are these orbital speeds so specific? Before I tell you the reasons, please let me address the remaining 3 variables first.

Meanwhile, if you forget everything I said in this section, don't forget the three orbital speeds I just listed.

5. The radius of celestial bodies: a master variable to respect if you are serious about finding the origin of the universe

"Mathematics compares the most diverse phenomena and discovers the secret analogies that unite them." — Joseph Fourier

The radius of the Earth, the Moon, and the Sun was known for thousands of years. For instance, Earth's radius has been known with extraordinary accuracy since about 240 BCE, when the Greek scholar Eratosthenes of Cyrene calculated it. Can you imagine that this African utilized shadows and geometry to estimate the Earth's size within 1% to 2% of the modern value? But that was exactly what he did. Isn't that amazing? It is not surprising, as Egypt (based in North Africa) used to be the center of excellence in the days when the Greeks ruled the world!

The ratio of the Moon's radius to the Earth's radius has been known since about 270–280 BCE, thanks to the work of the ancient Greek astronomer Aristarchus of Samos. But the precise dimension was calculated about 70 years ago.

Although estimates of the Sun's radius have been made for over 2,000 years, precise measurements were obtained in the late 19th and 20th centuries. Using modern space-based instruments, the Sun's radius has been known accurately for decades. As of today, here are the value of the radii of the celestial bodies I am interested in:

- Earth: 6,378.14 kilometers
- Moon: 1,738.10 kilometers
- Sun: 696,000 kilometers

Why is the Sun's radius more than 100 times that of the Earth, which is more than 3 times that of the Moon? Are these radii really caused by chance or random events, as the Big Bang theory says? How did these celestial bodies get their spherical shape? Please be patient with me until I reveal the shocking secret hidden beneath these trends, which scientists have not paid attention to for centuries!

The most important thing I want you to remember from this section is the radii of the 3 celestial bodies we are interested in.

6. If you think that the distance separating celestial bodies is by chance, think twice about what is called the semi-major axis

"Our thoughts make us what we are." — *Dale Carnegie*

What is the value of the distance separating the celestial bodies, and how can it help us understand the origin of the universe? Since the days of the Greek thinkers (more than 2000 years ago), efforts have been made to use geometry to calculate the distance between the Earth, Moon, and Sun. For instance, the first rational estimates (3rd Century BCE) of the Earth-Sun distance were done by Aristarchus of Samos, a Greek. However, the breakthrough occurred about 400 years ago. In fact, since Johannes Kepler published his laws of planetary motion between 1609 and 1619, the average distance between the Sun and the Earth has been understood in relative terms. Other scientists, including Sir Isaac Newton, also contributed to refining the theory that addresses this distance and the orbits of celestial bodies. Unfortunately, he linked those distances to what he thought were gravitational forces between celestial bodies.

For more than 70 years, the precise distance between the celestial bodies that interest me and the bodies they orbit is known:

- Mercury-Sun distance: 57,910,000 km (meaning the distance between the Sun and Mercury, the closest planet to the Sun)

- Earth-Sun distance: 149,600,000 km (meaning the distance between the Earth and the Sun)
- Moon-Earth distance: 384,400 km (that's the distance between the Moon and the Earth).

These distances are called the semi-major axes of Mercury, Earth, and the Moon, respectively. Why are the planets and satellites located at a specific distance from the celestial body they orbit? My discovery proved there is a story behind all these distances that scientists have labored to collect, but have failed to interpret properly. Yet, most theories you will read about the universe's origin don't even think about how those numbers are a foundation of cosmology.

All I want you to underline in this segment is the value of the Mercury-Sun distance, the Earth-Sun distance, and the Moon-Earth distance. Before I unveil how I discovered the story behind these distances, please let me address the last of the 4 variables first.

7. Understand this simple variable (escape velocity) discovered hundreds of years ago, and you will prove the origin of any celestial body in the universe easily

"The person who goes farthest is generally the one who is willing to do and dare." — Dale Carnegie

Last but definitely not least among the 4 variables, escape velocity was critical to me as I refined my theory of the origin of the universe. By the way, velocity is a fancy word physicists use to talk about the speed of an object moving in a certain direction. As its name indicates, escape velocity is the speed a body needs to get away from another body. Why is this? This is because most celestial bodies have gravity, which tries to "attract" or keep things on their surface. In general, the escape velocity is the speed needed for a body to escape the gravitational pull of another body.

Imagine this scenario, for instance. When you jump, you fall back on the surface of the Earth because of the Earth's gravity. But if you can jump with a speed equal to or superior to the Earth's escape velocity, you will not return to the Earth. Still, you will fly into space and never return, for that escape velocity won't allow the Earth ever to have any power to "attract or pull you back". In other words, the escape velocity is about the minimum speed required to overcome gravity.

As simple as this concept may sound, it has some unbelievable implications, but that has not crossed the mind of the scientists who, before me, dealt with the formation of the universe. Before telling you what I found, please give me a few minutes to review what is known about the escape velocity of the celestial bodies I am interested in.

The concept of escape velocity has been known since about 1700, when Isaac Newton formulated his law of universal gravitation. It was the practical application of this concept that led to the invention of rockets since the middle of the 20th century. For more than 200 years, the exact value of Earth's escape velocity was known, although the gravitational theory used to calculate it was wrong (See my book *Turbulent Origin of the Universe*).

As of today, here is the escape velocity of some of the celestial bodies I want to focus on:

- Sun: 617.6 km/s (that is approximately 55 times that of Earth).
- Earth: 11.186 km/s (that is about 7 miles per second or 25,000 mph)

Because I don't need the escape velocity of the Moon before proving my point, I did not bother listing its value here. You are free to forget everything I said about escape velocity. Still, please remember that the escape velocity is the minimum speed required to break free from a celestial body's gravity. In other words, the escape velocity of the Sun is the minimum speed required for a body to escape the gravity of the Sun. Likewise, Earth's escape velocity is the minimum speed required to escape Earth's gravity. Very soon, I will share with you what I discovered as the cosmological implication of the escape velocity. Now that I have introduced the 4 variables of interest, I will recap them for the 3 celestial bodies we will focus on to save time. That's what comes next.

8. He who can pay attention to these 12 simple facts collected over centuries and confirmed by NASA will crack the code of the universe properly. He who cannot will keep wasting time and money cosmologically

"The great danger we face is of being deceived by the successes and good fortune of our "engineering solutions" into thinking we really understand the "physics". But the real world has a way of shocking us back to reality when our "tried and tested" engineering model fails miserably on a completely new problem for which we have not calibrated it. This is what happens when we really don't understand the "physics" behind what we are doing." — William K. George

People have countless questions about the origin and formation of the universe, most of which nobody can answer completely. Likewise, scientific, philosophical theories, and religious beliefs propose countless answers or so-called answers to what, according to their perspective, they think are the most important cosmological questions. Sometimes, the problem we have in this world is not because of the answers people give, but because of the questions we ask. But are scientists really asking the right questions that will accurately lead us to the origin of the universe? What if a handful of questions could force scientific data to vomit up the answer we are looking for about the origin of the universe, thereby saving us time and money?

Thanks to my turbulence breakthrough, I discovered during my investigation of the timeline of the formation of the celestial bodies that any serious theory or study of the true origin of the universe must ask and answer scientifically the following 12 most important cosmological questions (that are based on scientific facts validated by NASA and other top space agencies across the globe):

1. Why is Mercury located at 57,910,000 km from the Sun?
2. Why is the distance separating the Earth and the Sun 149,600,000 km?
3. Why is the distance separating the Moon and the Earth 384,400 km?
4. Why is the orbital speed of the Sun 19.4 km/s?
5. Why is the Earth orbiting the Sun at 29.78 km/s?
6. Why is the Moon orbiting the Earth at 1.02 km/s?
7. Why is the radius of the Sun 696,000 km?
8. Why is the radius of the Earth 6,378.14 km?
9. Why is the radius of the Moon 1,738.1 km?
10. Why is the escape velocity of the Sun 617.6 km/s?
11. What is the escape velocity of the Earth, 11.186 km/s?
12. Which religion has a creation story matching these scientific evidences, and why is that?

I know that most people (even top scientists) reading this book have never seriously pondered any of these questions. Some of them are so trivial that if you ask them, some people may wonder why you are thinking about so simple stuff, and yet, as you will see very soon, that's the humble beginning of the wisest journey to crack the complex universe. Still, if you ask modern scientists to tell you a rational story that addresses all these questions, they could not, but they will most likely tell you that random events cause the numbers mentioned in them. In other words, if you now ask the smartest physicists who collect these data to answer the system of the above 12 questions, they cannot, except telling you that those numbers are due to chance or an accidental event. In fact, even the smartest physicists who have won the Nobel Prize never asked those questions carefully. Why? Because, while chasing other data they think are more important, most scientists ignore the facts related to these questions, and those who know the data mentioned in them disregard that they are the master key to unlock the code of the universe. Another reason is that it is easy to answer questions about the WHAT but not about the WHY and HOW. Hence, there are many scientific laws, but few clean theories explain the whys. In this book, you will discover the real WHY and HOW of those trends and many more.

Nathanael-Israel Israel! How do you know that these 12 questions are enough? You may continue to ask me. Well, now look. I may be wrong, and if I am, I want to be put right. Let's examine the facts first. Together, we are going to scrutinize the data scientifically and determine whether they are truly due to chance. I know some of you will look for me and thank me or try to collaborate with me in some projects, whether it is inviting me to speak at your organization (learn more at Science180.com/speaking), or interviewing me so I inspire and help your audience (learn more at Science180.com/interview, or investigating some questions together, or teaming up with me to achieve a goal dear to us (learn more at Science180.com/services). Without waiting any longer, so we can easily refer to them as we do some calculus later, let's recap in a table (Table 1) the data of the 4 variables and the celestial bodies that I used to formulate what I termed the 12 most important questions in cosmology.

Table 1: Summary of the 4 variables of the celestial bodies in the Solar System that I used to formulate the 12 most important questions in cosmology

Celestial bodies	Escape velocity (km/s)	Radius (km)	Semi-major axis (km)	Orbital speed (km/s)
Sun	617.6	696,000		19.4
Mercury			57,910,000	
Earth	11.186	6,378.14	149,600,000	29.78
Moon		1,738.10	384,400	1.02

Source of raw data: NASA (2018)

So far, I have explained the data I will use to build my theory. I also shared with you the 12 most important questions for cracking the code of the universe. Now that you have mastered the essential 4 variables and the questions, I will start using them to unravel the methodology I used in my research. If you are ready for it, let's go with some specifics.

TAKE-HOME MESSAGE OF SECTION 2: THE 4 SIMPLE SCIENTIFIC EVIDENCES REQUIRED TO CRACK THE ORIGIN OF THE UNIVERSE

In this chapter, I explained the 4 most important variables, according to my turbulence breakthrough, that are enough to demonstrate the origin of the universe and test the existence of God scientifically. These variables are the semi-major axis (i.e., the distance separating a celestial body from the one it orbits), the escape velocity (i.e., the minimum speed needed to escape the gravity of a celestial body), orbital speed (the speed at which a celestial body orbits another one), and the radius of the celestial bodies. In Table 1, I summarized the data coded in what I called the 12 most important questions in cosmology. We will refer to that table a lot as we engage in some calculus in the next chapter.

Section 3: Developing my Turbulent Split-Gathering Methodological Framework: the Foundation of my Inquiry

"Most of the important things in the world have been accomplished by people who have kept on trying when there seemed to be no hope at all." — Dale Carnegie

"Where success is concerned, people are not measured in inches, or pounds, or college degrees, or family background; they are measured by the size of their thinking" — David J. Schwartz

9. What am I going to tell you in this critical chapter?

"There ain't no rules around here. We're trying to accomplish something". — Thomas A. Edison

To properly explain the formation of the celestial bodies from the perspective of my turbulence breakthrough, you need to understand certain terms, vocabulary, concepts, rules, and logic I used to construct my theory. In this chapter, I will walk you through them.

Because my demonstration revolves around distance and speed, so that the duration of time could be calculated scientifically, I will explain the meanings of the distances and speeds I will use to understand how the formation of celestial bodies began and the scientific process that birthed them. Although some of the formulas may seem complicated, I will break them down into simple concepts that you will understand clearly. The chapter will explain the formula we will use to calculate the scientific date of birth of the celestial bodies. I will first clarify the forerunners of the celestial bodies, then describe the implications of the journey they took before placing them in their orbits and conferring on them their current characteristics. In the process, I will unveil all the formulas that I will use to demonstrate the origin of the universe. It is only after explaining my methodology that, in the next chapter, we can be ready to test whether any religious creation story agrees with scientific findings.

10. Do all celestial bodies have a common ancestor or precursor? Pay attention to what I called the "Turbulent Primary Materia," which is completely different from what Charles Darwin and evolutionists thought

"We should take care not to make the intellect our god; it has, of course, powerful muscles, but no personality." — Albert Einstein

The beginning of every human being, naturally born of a woman, can be traced back to the journey in their mother's womb. And to fully know how human beings are formed, you need to know what happens to babies inside their mother's womb. I have proven that all celestial bodies were at one point a baby that I called "precursor", which had to go through specific changes before becoming the adult celestial body known today. The precursor of a celestial body is like the direct forerunner, predecessor, forefather, antecedent, ancestor, pioneer, or trailblazer of that body.

How did I know that celestial bodies have a precursor? Because the math hidden behind the trends of the key variables collected on them said so. Which trends am I talking about? Some of these trends are related to the positioning, location, orbital shape, orientation, speed, rotation, mass, and size of the celestial bodies. Although every celestial body has a forerunner, before my discovery, scientists ignored how precursors formed and the processes by which they gave birth to the corresponding celestial bodies we know today.

Because my focus in this book is on the formation of the Earth, the Moon, and the Sun, I will focus my demonstration on the changes that their precursors went through, hoping that those who want to know the details of what happened to the precursors of the other bodies in the universe can refer to my book *Turbulent Origin of the Universe.*

The Solar System had a forerunner, which I called its precursor. The Sun itself had a precursor, which I called the Sun's precursor. Likewise, the precursor to the Earth-Moon system was its forerunner. The precursor of the Earth was the forerunner of the Earth that matured into the Earth. The precursor to the Moon was a body that changed to give birth to it.

But how did all these precursors of celestial bodies become the celestial bodies we know today? That's one of the greatest questions scientists have been trying to answer for centuries. Starting with the central process that controlled precursor formation, I will explain how these precursors were born and how they gave rise to the celestial bodies corresponding to them. The evidence presented in this book is rooted in a *Universe Turbulent Origin Formula* that I worked out while decoding the dynamics of the fluid in the precursors of celestial bodies, and then applied to calculate how long it took for the Sun, Moon, Earth, and more to form.

11. He who can master this simple concept of the "split-gathering of precursor of bodies" will have the key that Albert Einstein and Isaac Newton missed! He who cannot will suffer forever (eternal damnation?)

"Nothing is lost, nothing is created, everything is transformed." — Antoine Lavoisier

Just as, to fully know the process of the formation of human beings, it is very important to consider how they start with an egg that grows and passes through changes, so also to explain the formation of celestial bodies, it is imperative to know what their mother was and what happened to them inside the belly of such a mother. This may sound weird, but I have really shown that all celestial bodies descended from an initial matter, the mother of all particles that I called the *Turbulent Prima Materia.* This initial matter was not concentrated at a single point but was a bulk of original matter that covered a massive amount of space (which I cannot imagine) in the early universe.

SECTION 3: DEVELOPING MY METHODOLOGICAL FRAMEWORK: THE FOUNDATION OF MY INQUIRY

Just as the initial cells of all human beings are undetermined (meaning they have the power to become any cell), so also, I have shown that the initial matter in the universe had the potential to become any kind of matter. In biology, for instance, the process through which a living thing develops from an egg to become an adult is studied by following, over time and space, how the initial cell grows, multiplies, and differentiates into cells that form complex tissues, organs, appendages, and the full body.

Before my discovery, scientists had proposed various theories to explain how the universe formed, but they ignored how to twist the raw scientific data to craft properly those theories. While analyzing data on celestial bodies, I discovered that all the processes that shaped the initial matter in the universe and transformed it into the complex cosmos we see today can be explained by what I termed the "split-gathering" of precursors. In other words, the formation of all celestial bodies (including planets, satellites, stars, galaxies, and their clusters) can be boiled down to a simple process I termed "split-gathering" of body precursors. But what does "split-gathering" mean? What is the split-gathering of the precursor of bodies, and why is it so crucial?

As the name indicates, split-gathering involves two processes: splitting something into daughter bodies and gathering them into final bodies. Now that you know the definition of "split-gathering", my duty in the rest of this book is to give you a glance at the split-gathering process and how it shaped the formation of the Earth, the Moon, the Sun, and other celestial bodies in the universe. I will also use simple mathematics (that even laypeople can understand) to calculate the timeline of the formation of these bodies. Then, I will answer the most asked question of all time concerning the origin of the universe. By the way, whenever I used the verb "split-gather" (that I invented), please understand that I mean a process that splits and gathers certain precursors of bodies. The split-gathering followed specific rules. Just as you cannot understand the history of a child if you neglect the history of his or her parents during the pregnancy that preceded his or her birth, so also, I will first elucidate how the precursors of the celestial bodies were like pregnant women!

12. Understanding the pregnancy of the precursor of celestial bodies is very easy

"Making the decision to have a child is momentous. It is to decide forever to have your heart go walking around outside your body." — *Elizabeth Stone*

Most births followed a pregnancy. Although most people may support that a pregnancy can be experienced only by some living things, the analysis of the characteristics of the celestial bodies and the relationships between them suggested to me that their precursors went through what I labelled the pregnancy of precursors. In fact, just as a woman can become pregnant, and the egg will grow, pass through developmental stages (egg, embryo, fetus, baby), until the baby is born at a certain time (e.g. usually about 9 months for human beings), and then mature, passing through other life stages (e.g. baby, teenager, adolescent) before reaching an adult age, and die at some point (death being able to occur anytime), so also, the precursors of all celestial bodies were once like a gigantic "egg" inside the belly of their mother. Once each precursor was born, it became an adult celestial body. What was the nature of the mother of the celestial bodies, and how did it transition from its original stages to the final stages that yield the celestial bodies we see today? What was the nature of the needed changes?

In *Turbulent Origin of the Universe*, I proved that, at one point during their formation, the precursors of celestial bodies were fluid layers that moved at different speeds before splitting and coalescing into distinct bodies. I showed that everything started with a massive amount of initial matter that I called the "Turbulent Prima Materia," which appeared mysteriously in the universe. Some of you may want to know more about this initial particle. If you are interested in the details, please refer to *Turbulent Origin of the Universe*. What I want you to focus on is that the bulk of that initial matter was heavily destabilized, and that it started splitting and gathering into bodies following a cascade of breakup under the influence of turbulence.

The bodies that were born out of that split-gathering went through turbulent processes to become the adult celestial bodies we know today. Some of those changes in the precursors were caused by the turbulence they went through. Others were caused by the interactions between the precursors. Some were caused by the environment where those precursors matured. The size of the precursors, as well as the distance and the speed at which they traveled, affected the outcome of their bodies. Those who study the development of an embryo into an adult will tell you that, as the initial cell grows and matures, it differentiates into what are called differentiated cells, which, in normal conditions, cannot easily revert to the initial undifferentiated cells. Likewise, I discovered that, as the bulk of the initial matter underwent turbulent changes, complex structures formed within it and later became various types of chemical particles, the origin of which I developed in my book *Turbulent Origin of Chemical Particles*. Because of the scope of this book, I won't delve into those details here, but I need to bring them up so the curious can follow up if they want.

Today, all planets in the Solar System are spherical and located at specific distances. Most people don't even wonder what the machinery behind those trends is. Because distances and speeds (which I can use to calculate time) are what I want to use to show how celestial bodies were formed, of the many changes that these precursors of celestial bodies went through, I will focus just on those involving the distance they traveled, the speed of such travels, and how they got their spherical shape.

Before moving to the next segment, let me summarize what I just said. I just told you that, while studying the celestial bodies in the universe through the glance of the split-gathering framework, I discovered that, just as a seed can germinate, grow, branch out and develop into a big tree, and just as a single egg (of a living organism) can be born, grow, divide into various body parts having distinct and complex features, so also, the bulk of the initial matter in the universe split-gathered into daughter bodies, which went through processes that reorganized them into various precursors of bodies, which matured into the adult bodies we know today. So, where are we going next with all this? Well, I will next talk to you about how some precursors of bodies were organized into a primary body orbited by secondary bodies. Why is this important? Because to explain the journey of the precursor of bodies, I will need to refer to how the primary and secondary bodies were born. As I define these terms next, you can better understand what I mean.

13. The big secret of dealing with the birth of a primary body and its secondary bodies

"Sometimes the questions are complicated and the answers are simple." — *Dr. Seuss*

You certainly know that the Moon orbits the Earth, while the Earth orbits the Sun. In fact, besides the Sun, all other bodies in the Solar System orbit the Sun. But you probably never wondered why it is so? What if I tell you that, by using my turbulence breakthrough, I accurately pinpoint that all those movements started from somewhere? Before I give some details, I must first explain two simple terms that I will need later: primary bodies and secondary bodies.

Let me use the example of planets orbiting the Sun and of satellites orbiting their primary planet again. Considering the example I just gave (the Moon orbiting the Earth, while the Earth orbits the Sun), the body that is orbited by others is called a primary body. In contrast, the body that orbits the primary body is called a secondary body. For instance, the Sun is the primary body in the Solar System. Because planets orbit the Sun, they are the secondary bodies of the Sun. When you move to the scale of a planetary system (meaning the system formed by a planet and its satellites), the satellites become the secondary bodies, and the planet will be the primary body. In other words, when you consider the Earth-Moon system, the Moon is the secondary body while the Earth is the primary body. It is all about the reference. Please let this concept of reference sink in, because very soon, when we start talking about distance, speed, and duration, we will need to know the reference.

A system of celestial bodies is a network or an arrangement of a primary body and its secondary bodies. Hence, the Solar System is the arrangement of the Sun and all the bodies orbiting it. The Earth-Moon system is the system formed by the Earth and the Moon. Now that everybody understands what I meant by primary and secondary bodies in a system of celestial bodies, I will tell you what I discovered and that I detailed in *Turbulent Origin of the Universe.*

One thing people ignore is why some bodies orbit others. Today, some people take for granted that we know the Moon orbits the Earth and that the Earth orbits the Sun. But it took thousands of years of hard work for ancient and modern thinkers to discover those movements and the associated distances and speeds. At one point, when Galileo, for instance, was telling his contemporaries that it is the Earth that is turning around the Sun, not the reverse, he was mistreated and even jailed for life! Well, "whether Galileo was a huge jerk who mocked the Pope, even though he was supporting Galileo's research, or whether he was not jailed as much as he was put on house arrest in a comfortable house with visitors", please listen to me.

After spending 12 years working out the mathematics behind the massive data collected on the celestial bodies, that top scientists collected throughout the centuries, but that they discarded, while they keep going after new data using billions of dollars, I discovered that, in each system of bodies, the primary body and its secondary bodies once had forerunners, which I called the precursor of that system of bodies. It turned out that it was the precursor to the system of celestial bodies, which split into the precursor of the primary body and the system of secondary bodies (Figure 1). For instance, as the precursor of the Solar System was being formed, it birthed:

- The precursor of the Sun and
- The precursor of all the bodies orbiting the Sun.

In the same manner, as the precursor of the Earth-Moon system was split-gathering, it birthed:

- The precursor of the Earth, and
- The precursor of the Moon

Figure 1: Precursor of a system of bodies split-gathering into the precursor of primary and secondary bodies

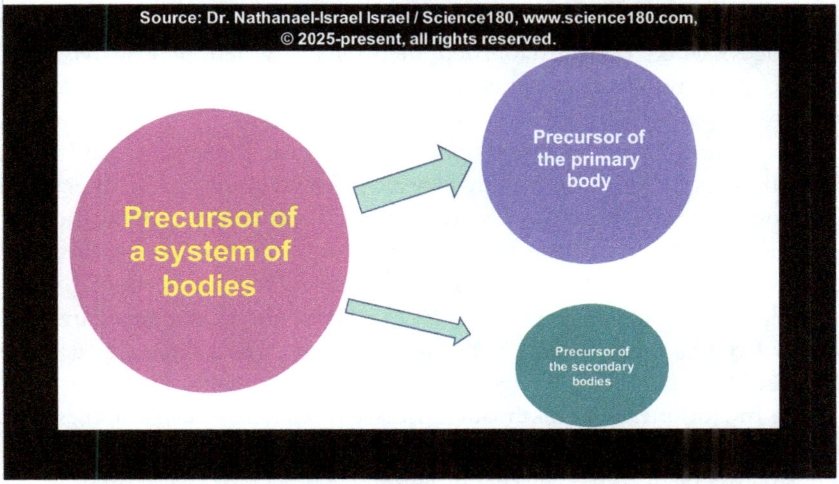

But the main problem is how and when those split-gatherings happened. Rather than repeating the lengthy demonstrations I did about that in my book *Turbulent Origin of the Universe*, I will now share with you some of the laws that explained the journey that led to the birth of these bodies. Once you understand what I am about to tell you, you will be in good shape to have the passkey to unlock the origin of the universe in your hands. Why am I saying that? Let's review together what I discovered, and you will understand! Please stay with me as I elaborate on the journey that the precursor of the celestial bodies had!

14. Can you ignore the journey of the precursors of the celestial bodies and still understand the universe's origin? No! Therefore, watch this!

"The only impossible journey is the one you never begin." — Tony Robbins

"A journey of thousand miles begins with a single step." — Loa Tsu

"Sometimes it is the journey that teaches you a lot about your destination." — Drake

"The important thing is not to stop questioning. Curiosity has its own reason for existing".
— Albert Einstein

Life is a journey. Before their birth, natural babies went through a journey in their mother's womb. Then, after spending time in that womb or incubator (in some cases), every human being was born on a certain date, called the date of birth, and then the life journey began. Some people moved from where they were born to other places, where they established themselves. When they arrive at their destination, they don't just stay there idly, doing nothing, but people move around to settle.

Although not "living", celestial bodies had a certain journey before arriving at their current orbital position. On their orbit, celestial bodies are neither idle, but move around other bodies (e.g., revolution) and sometimes around themselves (e.g., rotation). What was the initial journey of the precursor of the celestial bodies— and why on earth would you need to know it?

Thanks to my turbulence discovery, it turns out that understanding the journey of the celestial bodies' births is key to properly calculating when they formed and how long it took to reach their current positions.

Using my turbulence breakthrough, I discovered that celestial bodies did not reach their current positions or acquire their current shapes by chance. I realized that during the birth of the celestial bodies, their precursors were first:

- moved at a certain speed
- carried over a certain distance during a certain journey
- reached a certain point in the process and, then, were
- curled at a certain speed to get their form, or being gathered into a final (e.g., spherical) shape

For instance, the precursors of the Earth, the Moon, and the Sun traveled a long distance before undergoing the split-gathering process that rearranged their matter into spherical bodies. In other words, the precursor of each celestial body traveled a certain distance before reaching a certain position on orbit where its fluid layers were gathered into a spherical body. Before I tell you how I discovered the distance and speed involved in the journey of the precursors of celestial bodies, please allow me to simplify things by saying something very important. This thing is that, before celestial bodies were born, their precursors had a journey, which can be divided into 2 steps:

- A linear journey
- A spherical journey

Because I am the first person in history to discover the real journey of the precursors of the celestial bodies, I summarized it into two steps that I named after my name:

- **Nathanael-Israel Israel Linear Journey**: The journey from the moment that the precursor was born until it reaches its orbit or the position where it could be collected into a final body and

- **Nathanael-Israel Israel Gathering Journey:** The journey of that precursor from being long layers of matter (e.g., fluid layers) to a spherical body.

I used the adjective "Linear" to qualify the first component of this journey to express that during that process, the precursor of the celestial bodies moved as if it followed a straight linear path. I also used the word "Gathering" to qualify the second component of the journey, emphasizing that it was during this period that the precursor changed from being "Linear" fluid layers to being "Gathered" into a spherical body. Look at this as if you were carrying a rubber band or yarn over a certain distance at a certain speed until you reach a point, then curl the yarn or rubber all the way around. The duration of the whole process would be the time before the material (yarn or rubber) was curled, plus the time of the curling itself. Similarly, look at the precursors of the celestial bodies as fluid layers carried over a certain distance before being wrapped around at a certain speed. Very soon, I will explain to you what the distance and speed I am talking about are. But for now, before moving on to the next concept, please remember this:

I defined the "Nathanael-Israel Israel Turbulent Journey" (also called "Nathanael-Israel Israel Total Journey" of a celestial body as the journey of the formation of that celestial body, calculated by adding the linear journey and the gathering journeys:

Nathanael-Israel Israel Turbulent Journey = Nathanael-Israel Israel Linear Journey + Nathanael-Israel Israel Gathering Journey

(Equation 1)

In the next section, I will explain how I calculated each element involved in those journeys. But, beforehand, let me remind you of what I am trying to do with all this. I am showing you how I calculated the formation times of the celestial bodies. To get there, we need to understand the journey of their precursors so we can sum their components to determine the formation times of the celestial bodies. Once we know that scientifically, we can test if it matches any scientific or religious story about the existence of God. That being said, let's now review some details more carefully.

15. You cannot crack the origin of the universe if you ignore these 2 simple concepts about the journey of the forerunners of the celestial bodies

"It is good to have an end to journey toward, but it is the journey that matters, in the end." — Ursula K. Le Guin

"Focus on the journey, not the destination. Joy is found not in finishing an activity but in doing it." — Greg Anderson

"For in the true nature of things, if we rightly consider, every green tree is far more glorious than if it were made of gold and silver." — Martin Luther King Jr.

Here, I will explain the two time intervals required to calculate the dates of birth of celestial bodies. Each celestial body has a specific location and shape. In the Solar System, for instance, celestial bodies have their own orbits (the trajectories along which they move) and specific shapes. For instance, the Earth, the Moon, and the Sun are spherical and each moves along its orbit. During my research into the origins of the celestial bodies, I showed that their precursors began their journey from somewhere before reaching their orbits. Similarly, their precursors were not born spherical; it took some processes for their constituents, matter, or particles to be collected into spherical bodies. In short, it took some time before the precursor of the celestial bodies reached their orbits, and it took another time for them to collect their matter into the spherical form that we know them today. Therefore, to calculate the duration (T) of the formation of the celestial bodies, we need two timings:

- **Nathanael-Israel Israel Linear Time** (T1): the duration of time it took for their precursor to move from where they were born to the position where they were collected together and put into orbit, and
- **Nathanael-Israel Israel Gathering Time (**T2): the duration of time it took for these precursors to be collected together into the spherical bodies we know today.

In other words, the Nathanael-Israel Israel Linear Time is the duration of what I called the Nathanael-Israel Israel Linear Journey. In contrast, the Nathanael-Israel Israel Gathering Time is the duration of the Nathanael-Israel Israel Gathering Journey. I purposefully named these times after my name because I am the first person to demonstrate how to use them to calculate the durations of the formation of the celestial bodies. Very soon, you will see their significance and why I need to label those journeys, speeds, and distances properly.

The duration of the complete formation of the celestial bodies is what I termed the Nathanael-Israel Israel Birthdate of the celestial bodies, which is the sum of the Nathanael-Israel Israel Linear Time and the Nathanael-Israel Israel Gathering Time. In other words, the Nathanael-Israel Israel Birthdate (T) of the celestial bodies is:

$$T = T1 + T2$$

SECTION 3: DEVELOPING MY METHODOLOGICAL FRAMEWORK: THE FOUNDATION OF MY INQUIRY

Nathanael-Israel Israel Birthdate = Nathanael-Israel Israel Linear Time + Nathanael-Israel Israel Gathering Time

(Equation 2)

The puzzling thing that all scientists before me have ignored is how to calculate these time intervals. In fact, breaking down the history of the birthdates of the celestial bodies in the aforementioned simple two journeys alone is a breakthrough, and calculating the duration of these journeys is even a greater challenge we need to overcome. Therefore, the main thing I will handle in the rest of this book is to clearly explain to you how I calculate each of the two components of the duration of the formation of the celestial bodies. Then we will use the number we got to see if it matches any story available in the literature.

To make it easier for you to follow, let's first map out how I want to calculate the 2 kinds of timing I mentioned above.

Imagine you took a journey using a vehicle. As you know, to calculate the duration of your journey, you need to know:

- the length of the journey and
- the speed of your travel.

Then, you divide the length of the journey by the speed to obtain the duration of time. I will also use the same simple approach to calculate the two kinds of timing in the journey of the precursor of the celestial bodies. I will first generalize the formula, then apply it to the Earth, Moon, and Sun. Although the formulas in this book may sound complex, they are all based on the simple math I just mentioned, which can be summarized as dividing a distance by a speed to calculate the duration of time.

Likewise, for each of these durations of time (Linear time versus Gathering time), once we know the distance traveled and the speed of the travel, we will divide each distance by its corresponding speed to get the duration of time. Then, we will add them up (Linear time + Gathering time). It is simple math. If you have already browsed this book and see some big formulas that scared you, don't worry. They are very simple, and even a 9-year-old kid can fully understand them.

However, as simple as this mathematics may sound, you will be shocked that, for thousands of years, it never crossed the mind of any scientist. Why? Although specific distances separate celestial bodies and they move at specific speeds, people don't know the history of these distances and speeds. By the end of this book, I will share with you what scientists think the journey of the precursors of celestial bodies was, and how their mistaken references led them to miss the proper understanding of the origin of the universe. For the sake of time and space, I will not show the entire mathematical demonstration of how I discovered these stories here, as I already presented that in *Turbulent Origin of the Universe*; instead, I will summarize what you need to know quickly. Remember! This book is the bottom line of my discovery, and I want everybody, even the laypeople who never set foot in any middle school, to fully understand it, provided they know how to divide a number by another. That's all I need from you. Using a calculator (maybe from your phone), you can also check the math with me in the next chapter and get the answer by yourself. That's it. I guarantee you are going to laugh aloud when I show you this simple math that Albert Einstein and Sir Isaac Newton missed, and "plunged" modern science into an abyss scientists have been ignoring for more than 100 years! Then, you can claim your freedom from scientistic and religious nonsensical theories that try to suffocate most people, maybe including you, more than they (maybe you) think. For now, hold on and bear with me for a while. I am telling you the truth. I know what I found is historic, and the whole world will be talking about it very soon (if not already by the time this reaches you)!

For now, let me quickly handle the calculation of the "Nathanael-Israel Israel Linear Time" (T1), meaning the time it took for the precursor of celestial bodies to move from where they were born to the position where they were ready to be collected or gathered together into a spherical body and put into orbit. After I calculate that linear time, I will turn to "Nathanael-Israel Israel Gathering Time" (T2), the duration it took for these precursors to be collected together into the spherical bodies we know today. Once the linear and gathering times are known, we will add them together and ring the bell for the math of the celestial bodies' birthdates. Then, we will see if it can prove the existence of God scientifically. If that sounds good, let's go!

15.1. Accept the Nathanael-Israel Israel Linear Time (T1), and it will work wonders for you

I told you in the previous section that 2 types of time durations are crucial for understanding the formation of the universe. Here, I will tackle the first one: the "Nathanael-Israel Israel Linear Time". I hope it won't hurt if I remind you that this is the duration of the precursor's journey from where they were born to where they were gathered together and put into orbit. To refresh our minds, let's recall that the calculation of this duration of time necessitates the knowledge of:

- The distance traveled by the precursor during the linear journey (that journey is what I called the "Nathanael-Israel Israel Linear Journey"), and
- The speed of the journey.

Because I invented the methodology I am explaining here, I coined the terms "Nathanael-Israel Israel Linear Distance" and "Nathanael-Israel Israel Linear Speed" to describe the distance and speed I just mentioned for the "Nathanael-Israel Israel Linear Journey". In other words, during the "Nathanael-Israel Israel Journey", the precursor of the celestial bodies traveled a "linear" distance that I termed the "Nathanael-Israel Israel Linear Distance" at a speed that I named the "Nathanael-Israel Israel Linear Speed". For instance, the precursors of the Earth, the Moon, and the Sun did not just randomly fly from somewhere and magically or miraculously become them. Before defining the formula of the "Nathanael-Israel Israel Linear Time" mathematically, I first need to walk you through 3 things:

- How the fluid layers of the precursors were organized as layers (that are the precursors of some strata in celestial bodies today)
- What the exact value of the "Nathanael-Israel Israel Linear Distance" is
- What the exact value of the "Nathanael-Israel Israel Linear Speed" is

Once we know these 3 things, the math is easy. We will just divide the distance by the speed.

15.1.1 If you deny that the precursors of celestial bodies were organized in layers, you are headed for scientific trouble

The trends in the orbital speeds and other variables of the celestial bodies showed that the bodies' precursor fluid layers (Figure 2) were not only organized into layers but also moved at different speeds, with the top layers moving faster than the layers beneath them (see details in *Turbulent Origin of the Universe*).

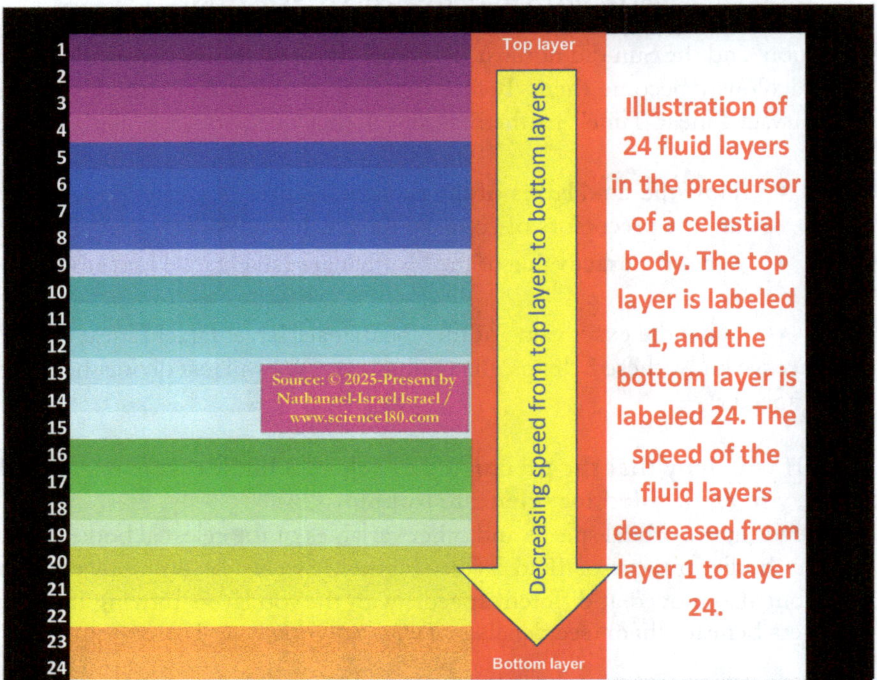

Figure 2: Fluid layers in the precursor of a body or precursor of a system of bodies

In other words, one of the key data points that helped me discover the organization of the matter in the precursors is the trends in the orbital speeds and orientations of the celestial bodies as a function of their distance from their primary bodies. During their movement, the layers separated from one another and were collected in bodies.

Using the split-gathering model that I developed, I showed that the precursors of celestial bodies were organized in a stack of fluid layers. I proved that the existence of these layers in the precursor of the Earth, for instance, is what caused the upper part of the Earth's crust to appear as strata or layers when the Earth is dug or excavated to build roads (Figure 3).

Figure 3: Roadside showing some of Earth's layers formed as a consequence of the fluid layers in Earth's precursor

Some upper layers of the Earth were caused by events (floods, erosion, sedimentation, etc.) that occurred on Earth after the formation of the universe. I wish I have room here to talk about the presence, age, and source of the fossils in Earth layers. I dealt with that in other books including *Turbulent Origin of the Universe.*

So I don't lose some people to overwhelming, unnecessary data, I will not go over all those technical details here, but I will elaborate on how the fluid layers of the precursor bodies orbiting the Sun were positioned.

Because Mercury (the closest planet to the Sun) is the main innermost body orbiting the Sun, I showed that the fluid layers of the precursor of Mercury were positioned on top of all the fluid layers of the bodies orbiting the Sun. I realized that the precursor of the bodies orbiting the Sun was once organized as fluid layers, which embedded the fluid layers of the precursor of the Earth-Moon system, and moved at a certain speed (which I will tell you very soon) away from the precursor of the Sun. In other words, as the fluid layers of the bodies orbiting the Sun escaped the precursor of the Sun, the fluid layers of the precursor of the Earth-Moon were positioned somewhere beneath the fluid layers of the precursor of Mercury, Venus, and all the asteroids located between the Earth and the Sun. Beneath the fluid layers of the precursor to the Earth-Moon system, other fluid layers existed. I will get back to that later.

When these fluid layers started splitting, the top ones split before the bottom ones, and, in the end, the organization of bodies around their primary bodies is caused by the story of the split-gathering of the fluid layers of the precursors of these bodies. Hence, the fluid layers of the closest bodies to the Sun were the first fluid layers to split from the stack of fluid layers of the precursor of the bodies orbiting the Sun. In other words, because Mercury is the closest planet to the Sun, the fluid layers of its precursor were the first layers to split from the precursor of the bodies orbiting the Sun (Figure 4). Now that I have explained my perspective on the organization of the precursor fluid layers, I will present how they split.

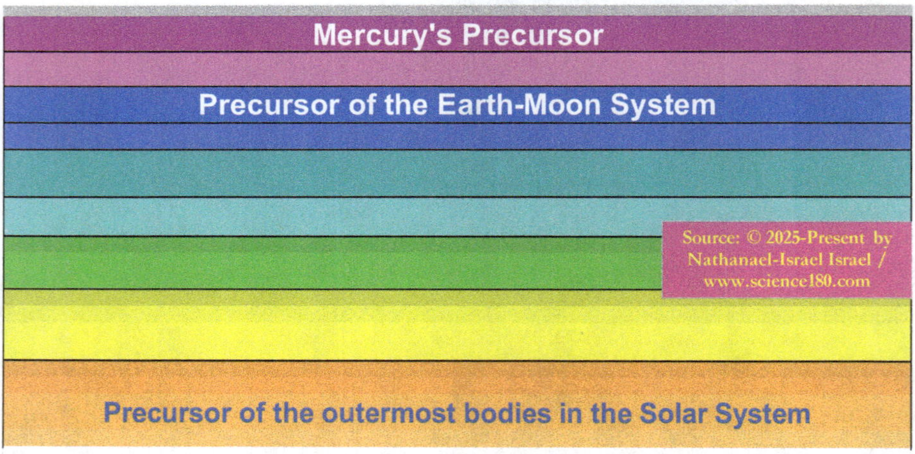

Figure 4: Fluid layers of Mercury on top of the fluid layers of the precursor of the bodies orbiting the Sun. Due to its position, it would be the first to split from the stack

15.1.2. The precursor of the secondary bodies escaped the precursor of their primary body

Every birth implies something coming out of something. For instance, when babies are born, they come out of their mother's womb. In the same way, the precursors of celestial bodies birthed their daughters not by going into a travail, but by pouring out their fluids into them in a specific way. The turbulent processes that shook the precursors and set them in motion explained how the split occurred. Because the precursors of the celestial bodies were moving when they formed, their fluids were moved in a special way, and during their separation, their daughter bodies were born.

In short, I proved that during the split-gathering of the precursor of celestial bodies, the fluid layers of the precursor of the primary body escaped the fluid of the secondary bodies. Imagine this as how a certain volume of fluid can be thrown into the air and split into clusters. For instance, the precursor of the Solar System was formed by allowing the fluid of the precursor bodies orbiting the Sun to come out of the fluid of the precursor of the Sun. In the same manner, I showed that when the precursor of the Earth-Moon system was birthing its daughter bodies, the precursor of the Moon escaped the precursor of the Earth.

Knowing that the precursor of the celestial bodies escaped other precursors is not yet sufficient to calculate the "Nathanael-Israel Israel Linear Time". We also need to know the speed of that escape and the length of the "Nathanael-Israel Israel Linear Journey". Before I address the length of the journey, I will first elaborate on the speed of the escape.

15.1.3. Want to know how fast the precursors of celestial bodies get to their current position in orbit? Then, master the Nathanael-Israel Israel Linear Speed

The "Nathanael-Israel Israel Linear Speed" of a body is the speed at which the precursor of that body escaped the precursor of its primary body and traveled the "Nathanael-Israel Israel Linear Distance". As I worked through the scientific data, I found that the "Nathanael-Israel Israel Linear Speed" is the speed that physicists call the escape velocity, which I already explained as one of the 4 most important variables (see chapter 2).

As a general rule, I showed that the fluid layers of the precursor of the secondary bodies escaped the precursor of their primary body at about the escape velocity of the primary body. But for more than 400 years, since the days of Sir Isaac Newton, no scientist before me really knew the significance of the escape velocity and how to use it to calculate the birthdate of the celestial bodies. All they did with their understanding of the escape velocity concept that Sir Isaac Newton invented was use it to build rockets and launch spacecraft into space, with the first man escaping Earth's gravity occurring about 70 years ago.

Scientists and engineers have not properly wondered about what if the precursor of the Moon could have escaped the precursor of the Earth, and what if the precursor of the Earth could have escaped the precursor of the Sun. Instead, until the publication of my discovery in 2025, scientists thought that the planets' journeys in the Solar System began somewhere else I will clearly tell you before ending this book. Because I don't want to put the chariot before the cow, for now, I will ask you to please stay tuned! Later in this book, I will also explain the consequences of this poor cosmological assumption for the scientific community. As you read *Turbulent Origin of the Universe*, you will see how, without a prior assumption, but using induction—meaning a logical reasoning based on collections of empirical observations, which, when patterns within observations emerge, can inform more generalized theory formation—and properly using deduction based on sound axioms, I discovered the correct path of the journey of the precursor of celestial bodies and the distance and speed of their journey. Those who attend Science180 Academy (www.Science180Academy.com) can also get a quick explanation of my detailed method and get their questions answered quickly. You can also invite me to speak at your organization. For now, let me help you to understand this critical matter quickly.

Applying the concept of the "Nathanael-Israel Israel Linear Speed" to the formation of the Solar System, the precursor of the bodies orbiting the Sun escaped the precursor of the Sun at about the escape velocity of the Sun. Because the precursor of the Earth-Moon System was embedded in the precursor of the bodies orbiting the Sun, the precursor of the Earth-Moon System escaped the precursor of the Sun at the escape velocity of the Sun. Likewise, as the precursor of the Earth-Moon birthed its daughter bodies, the precursor of the Moon escaped the precursor of the Earth at about the Earth's escape velocity. In other words, the "Nathanael-Israel Israel Linear Speed" of the Earth is the escape velocity of the Sun. The "Nathanael-Israel Israel Linear Speed" of the Moon is the Earth's escape velocity. If you forget everything in this segment, please remember this:

Nathanael-Israel Israel Linear Speed of the Earth = Sun's Escape Velocity

(Equation 3)

Nathanael-Israel Israel Linear Speed of the Moon = Earth's Escape Velocity

(Equation 4)

Nathanael-Israel Israel Linear Speed of the Mercury = Mercury's Escape Velocity

(Equation 5)

I will later tell you the unique importance of the "Nathanael-Israel Israel Linear Speed" of the Mercury. For those who may have forgotten what the "Nathanael-Israel Israel Linear Speed" means, let me recall that it is the speed at which the precursor (or forerunner) of these bodies moved from the moment they were born until reaching the position where they would be gathered together into spherical bodies. If you forgot why we need this speed, it is so we can use it as the denominator of the formula to calculate the duration of the Nathanael-Israel Israel Linear Time.

15.1.4. Looking for an easy way to know why celestial bodies are located at specific places? Then, don't ignore the Nathanael-Israel Israel Linear Distance

After escaping the precursors of their primary bodies, the precursors of the secondary bodies traveled a certain distance before arriving at the proper position, where they were collected into their final shape. That distance was required for the precursor fluid to be recognized as structures that, among many things, triggered its separation from the layers beneath it. Using data collected and/or verified by NASA and other agencies on the celestial bodies, I found that the precursor of the secondary body traveled about the distance separating the primary body from the corresponding secondary body. Applying this concept to the formation of the Solar System, I showed that, after the stack of fluid layers of the precursor of the bodies orbiting the Sun escaped the precursor of the Sun, it traveled about the distance separating the Sun and the Earth before the precursor of the Earth-Moon system separated from that stack. I proved that, after splitting from the aforementioned fluid stack, the precursor of the Earth-Moon system gave birth to the precursors of the Earth and the Moon. Then, the precursor of the Moon traveled about the distance between the Earth and the Moon before reaching the position from which it was collected into the Moon.

In other words, the "Nathanael-Israel Israel Linear Distance" of the Earth is the distance between the Sun and the Earth, which in Table 1 is called the Earth's semi-major axis. In contrast, the "Nathanael-Israel Israel Linear Distance" of the Moon is the distance separating the Moon and the Earth. The average distance separating the Moon and the Earth is called the semi-major axis of the Moon, while the average distance separating the Earth and the Sun is called the semi-major axis of the Earth (see Table 1).

In physics, it is not by chance that distance and time are calculated using specific reference frames. For instance, the "Nathanael-Israel Israel Linear Distance" I just calculated for the Moon was calculated with respect to the Earth's position. The distance between the Moon and the Earth is measured relative to the Earth. Likewise, the "Nathanael-Israel Israel Linear Distance" of the Earth is with respect to the Sun. For now, let's recap the "Nathanael-Israel Israel Linear Distance" in a way that we can remember quickly later:

Nathanael-Israel Israel Linear Distance of the Earth = Earth-Sun Distance

(Equation 6)

with Earth-Sun Distance being the distance separating the Earth and the Sun.

Nathanael-Israel Israel Linear Distance of the Moon = Moon-Earth Distance

(Equation 7)

with Moon-Earth Distance being the distance separating the Moon and the Earth.

Nathanael-Israel Israel Linear Distance of Mercury = Mercury-Sun Distance

(Equation 8)

with Mercury-Sun Distance being the distance separating Mercury and the Sun.

The main thing I want you to remember is that the formulas above show the distance traveled by the precursor of each of these celestial bodies before reaching a position where their fluid layers were ready to be collected into a spherical body. If you want to know the raw math behind these formulas, please refer to *Turbulent Origin of the Universe.* Now that we know the "Nathanael-Israel Israel Linear Distance" and the "Nathanael-Israel Israel Linear Speed", let's finally calculate the "Nathanael-Israel Israel Linear Time". We are getting close to the end. Figure 5 recaps how the precursor of a secondary body escaped the precursor of its primary body

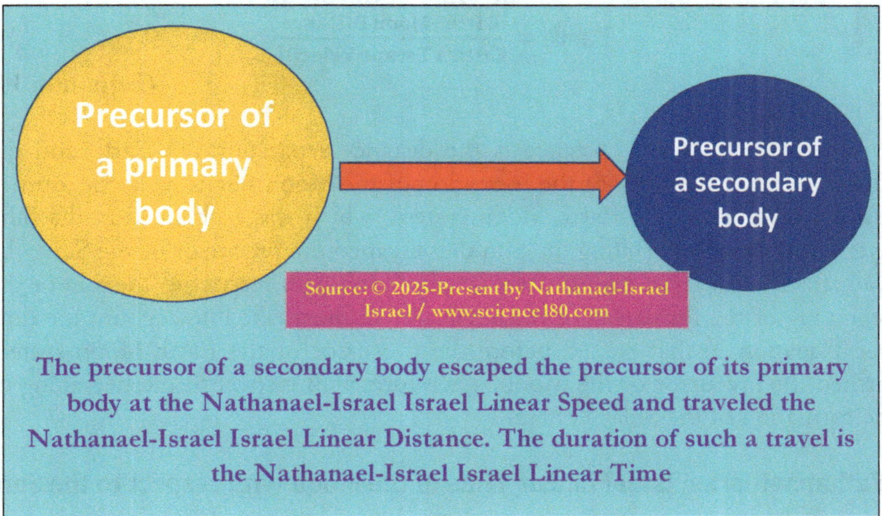

Figure 5: Illustration of the escape velocity, the speed at which the precursor of a secondary body escaped the precursor of its primary body. For instance, the precursor of the satellites escaped the precursor of their primary planet at the Nathanael-Israel Israel Linear Speed

15.1.5. Nathanael-Israel Israel Linear Time of the Earth, the Moon, and the Sun

First, please allow me to recapitulate what we have learned so far about the "Nathanael-Israel Israel Linear Time." I showed you that before the celestial bodies were formed, they escaped the precursor of their primary body at a certain speed termed the "Nathanael-Israel Israel Linear Speed" and traveled a certain distance that I titled "Nathanael-Israel Israel Linear Distance" before their fluid layers were ready to be wrapped around to collect together into spherical bodies. Considering the equations I developed for this linear speed and distance, the "Nathanael-Israel Israel Linear Time" of the Earth, Moon, and Sun can be summarized as:

$$\text{Nathanael-Israel Israel Linear Time of the Earth} = \frac{\text{Earth–Sun Distance}}{\text{Sun's Escape Velocity}} \quad \text{(Equation 9)}$$

with Earth-Sun Distance = Distance separating the Earth and the Sun.

$$\textbf{Nathanael-Israel Israel Linear Time of the Moon with respect to the Earth} = \frac{\text{Earth}-\text{Moon Distance}}{\text{Earth's Escape Velocity}}$$

(Equation 10)

where Earth-Moon Distance is the distance separating the Earth and the Moon. Because the story of the formation of the Moon started with the journey of the precursor of the Earth-Moon system (which was embedded in the fluid layers of the bodies orbiting the Sun that escaped the precursor of the Sun), the Nathanael-Israel Israel Linear Time of the Moon with respect to the Sun or the beginning of the formation of the Solar System must take into account the time (see Equation 9) that passed before the precursor of the Earth-Moon system birthed the precursor of the Moon. Therefore, with respect to the beginning of the formation of the Solar System:

$$\textbf{Nathanael-Israel Israel Linear Time of the Moon with respect to the Sun} = \frac{\text{Earth}-\text{Sun Distance}}{\text{Sun's Escape Velocity}} + \frac{\text{Earth}-\text{Moon Distance}}{\text{Earth's Escape Velocity}}$$

(Equation 11)

Now that I showed the expression of the "Nathanael-Israel Israel Linear Time" of the Earth and the Moon, you may be wondering what about that of the Sun. I purposely left that out until now. Why?

Indeed, to calculate the "Nathanael-Israel Israel Linear Time" of the Sun, I need to first explain to you the exact date that the precursor of the Sun started collecting into a spherical body. I first need to explain when the precursor of the Sun's innermost body formed. Then the timing of the precursor to the Sun's gathering will make more sense. Although it is said that the Sun (let's say the Solar System) is orbiting the Milky Way Galaxy (the galaxy we belong to), I did not calculate the "Nathanael-Israel Israel Linear Time" of the Sun by referring to the distance separating the Sun and the center of our Galaxy. The precursor of the Solar System did NOT descend from a point located at the center of our Galaxy. Even the precursor to the Solar System was not a single point but a massive body of matter, whose volume need not be known to demonstrate what I intend to do. Therefore, due to the nature of this bottom-line book, I will not dwell much on this aspect of my demonstration here, but those who want can consult *Turbulent Origin of the Universe*. Therefore, I used other details inside the Solar System itself to demonstrate the formation of the Sun.

I showed in *Turbulent Origin of the Universe* that the time elapsed before the precursor of the Sun was fully formed and ready to start gathering its fluid layers into a spherical body is about the time it took for all the fluid layers of the precursor of the bodies orbiting the Sun to move out completely. Using data on the orbital speed trends of bodies orbiting the Sun, I proved that the fluid layers of the precursor of Mercury, the closest planet to the Sun, were located above the fluid layers of the precursors of all the bodies orbiting the Sun. I showed that the duration of the escape of the precursor of all the bodies orbiting the Sun was less than the time it took for the stack of fluid layers of the bodies orbiting the Sun to move from the precursor of the Sun to about the position of Mercury. Because the precursor of the bodies orbiting the Sun escaped the precursor of the Sun at about the escape velocity of the Sun, the duration of this travel is about "Mercury-Sun Distance", the distance separating Mercury and the Sun divided by the escape velocity of the Sun. That duration is what I termed the "Nathanael-Israel Israel Linear Time" of Mercury:

$$\textbf{Nathanael-Israel Israel Linear Time of Mercury} = \frac{\text{Earth}-\text{Mercury Distance}}{\text{Sun's Escape Velocity}}$$

(Equation 12)

Therefore, considering what I said,

$$\textbf{Nathanael-Israel Israel Linear Time of the Sun} = \frac{\text{Mercury}-\text{Sun Distance}}{\text{Sun's Escape Velocity}}$$

(Equation 13)

If you are wondering where we are going with all these formulas, please let me recap. We said earlier that the journey of the precursor of the celestial bodies can be summarized into 2: a linear journey and a gathering journey. We just finished dealing with the equations of the linear journey, meaning the journey from the moment the precursor was born to the moment they are gathered into spherical bodies. We are getting close to the end of the tunnel. If you use the data in Table 1 and the formula above, you can already estimate the "Nathanael-Israel Linear Time" of the Earth, the Moon, and the Sun. But before we start plugging those numbers in, let's first scrutinize the second part of the equation of the date of birth of the celestial bodies.

Now that you know the ingredients of the first component of the formula of the duration of the formation of the celestial bodies we are interested in, I will next turn to the second component: the "Nathanael-Israel Israel Gathering Time", meaning the time it took for the celestial bodies to wrap their fluid around to form spherical bodies. Once that formula is developed, I will use the data in Table 1 to determine the exact duration of the celestial bodies' journeys. Please stick with me. We are almost there.

15.2. If you want to know accurately how long it took to form the celestial bodies, don't kick over the Nathanael-Israel Israel Gathering Time

As I explained earlier, the "Nathanael-Israel Israel Gathering Time" is about the duration the fluid layers of the precursor of the celestial bodies took to collect into the final shape corresponding to these celestial bodies. Why is this gathering time important? Because, to calculate the duration of the formation of the celestial bodies from the perspective of my turbulence breakthrough, we need to consider both:

- The time it took for the precursors to journey from the position where they were born all the way to the position where the fluid layers of the precursors arrived on the position of orbit where they were ready to be gathered together and
- The time it took for the fluid layers to collect into their final shape (e.g., spherical) once they reached the position or location of their orbit.

In the previous segment, I told you that the precursors of celestial bodies traveled a certain distance, which I named the "Nathanael-Israel Israel Linear Distance," at a speed I termed the "Nathanael-Israel Israel Linear Speed" before arriving at the position where they were collected into their final shape. And I showed you how to calculate the "Nathanael-Israel Israel Linear Time", which is the duration of that linear journey. Because the "Nathanael-Israel Israel Linear Time" is about the amount of time that passed before the final gathering started, the events I will discuss concerning the "Nathanael-Israel Israel Gathering Time" happened after the precursor of the celestial bodies reached about the orbit of the bodies.

After explaining how you can calculate the duration of the process that molded the fluid layers of the precursor into their final shape we know today, you will have in hand the elements of the mathematical formula needed to calculate the timeline of the formation of the celestial bodies. Because the Earth, the Moon, and the Sun, which I am interested in, are all spherical, the gathering time I will be dealing with here is the duration of the processes that caused their fluid layers to transition from their linear form to the spherical shape of these bodies.

When I was working out the mathematics of the formation of the universe, it turned out that the gigantic stack of matter in the precursors of the Earth, the Moon, and the Sun was once curled or wrapped around, giving these celestial bodies a spherical shape. Just before that, their precursors had a specific length and curled around at a specific speed. Therefore, to calculate the duration of the gathering of these precursors precisely, you need to know two very important things, which, to standardize my terminology, I expressed using two terms I coined after the name of their discoverer:

- "Nathanael-Israel Israel Gathering Distance": The length of the precursor of the celestial bodies just before their wrapping around started, and

- "Nathanael-Israel Israel Gathering Speed": The speed of the wrapping around of the precursor's fluid layers

Once we know these two components, we can easily calculate the "Nathanael-Israel Israel Gathering Time" (i.e., the duration of wrapping or gathering) by dividing the length of the precursor's fluid layers by the wrapping speed. I hope you remember our formula that relates the duration to the distance traveled and the speed. I will first discuss the formula for the length of the precursor of the celestial bodies just before the start of their wrapping around.

15.2.1 If you want to know how celestial bodies got their spherical shape, start with the Nathanael-Israel Israel Gathering Distance

Before my discovery, scientists thought that random events just gathered matter to form celestial bodies. But I proved that celestial bodies were composed of certain constituents and organized into a sphere according to certain rules. For instance, I just told you that, based on my investigation, the fluid layers of the precursor of the Earth, Moon, and Sun curled, swirled, or wrapped around to collect into their spherical form. If we can know the length of these fluid layers and the speed at which they were wrapped, we can easily determine the time it took to gather them in spherical bodies by dividing the length of the fluid layers by the speed of their wrapping around. But how can we know or calculate the length of these fluid layers just before they were gathered into spherical bodies? That was one of the breakthroughs I also made in turbulence.

Imagine a crochet yarn or a thread having a length *D* that you want to wrap all around a stick, whose circumference is the length of the yarn. Or imagine you want to wrap that thread all the way around to form a circle. Knowing the speed of the rolling around and the length of the crochet yarn, you can calculate the duration of the rolling by dividing the length of the yarn by the speed of the rolling. Likewise, to know the duration of that wrapping around of the precursor of the celestial bodies, you need to know the length of their thread and the speed of their curling or wrapping around.

To be sure everybody got what I said, let me use another example you can experiment with yourself. If you stand by a river and look at its water flowing, you will observe that some water filaments moving along the current curl around to form spherical or near-spherical 3D bodies called water vortices (singular = vortex). In physics, the study of vortices is a big deal for those who investigate turbulence, the high-level subject I had to handle before cracking the code of the universe's formation. If you Google the equations used to study turbulence and vortices, you will be shocked by how long and complicated they are.

Using my turbulence breakthrough (details in *Turbulent Origin of the Universe*), I discovered that, just before the fluid layers of the precursors of the celestial bodies were gathered together, they were as long as the circumference of the celestial bodies they became. For instance, if you consider the precursor to Earth, just before its fluid layers were gathered together to form the spherical Earth, they were as long as the Earth's circumference. Therefore, in that book, I used the adjective "circumferential", which means something related to a circumference, to express the gathering together of the fluid layers of the precursor of the bodies. In other words, the "Nathanael-Israel Israel Gathering Time" of a celestial body is the duration of the processes involved in the collecting together or the swirling or wrapping around of the fluid layers of the precursor of that celestial body into its spherical shape once that precursor reaches the orbit and is long as the circumference of the bodies they will birth.

I proved that, just before the circumferential gathering of the precursor of a celestial body, its fluid layers were as long as the circumference of the celestial body formed from them. Using the radius (R) of a circle, the circumference of a circle can be calculated by multiplying the diameter (which is twice the radius) by the scientific number called pi (which is about 3.14).

Circumference = 2 x 3.14 x Radius

(Equation 14)

Using the above formula, you can calculate the circumference of the Earth, the Moon, and the Sun. Why are we talking about this? To determine the duration of the gathering of the precursor of the celestial bodies once they reach their orbit, we need to know the length of their fluid layers, so we can then divide it by the speed of their wrapping.

Imagine you carried a long spaghetti noodle from your kitchen to your dining room and wrapped it around your fork. The time that experience took should include the time it took to travel from the kitchen to the dining room and the time it took to wrap the spaghetti around once you arrive in the dining room. Comparing the fluid layers of the precursor of the celestial as a long spaghetti noodle, the "Nathanael-Israel Israel Linear Time" is like the time it took to carry the spaghetti from the kitchen to the dining room, and the "Nathanael-Israel Israel Gathering Time" is like the time it took to wrap the spaghetti around the fork in the dining room. The "Nathanael-Israel Israel Gathering Distance" we are talking about here is like the length of a long spaghetti.

What I just told you is that the circumference of the celestial bodies is about the length of their fluid layers just before they curled to form these celestial bodies. The length of those layers is what I called the "Nathanael-Israel Israel Gathering Distance". Therefore, considering Equation 14, we can say that:

Nathanael-Israel Israel Gathering Distance of the Earth = 2 x 3.14 x Earth's Radius

(Equation 15)

Nathanael-Israel Israel Gathering Distance of the Moon = 2 x 3.14 x Moon's Radius

(Equation 16)

Nathanael-Israel Israel Gathering Distance of the Sun = 2 x 3.14 x Sun's Radius

(Equation 17)

Where are we in our logic, and how useful is this math? Well, before I continue my story, let me remind you where we are then. To determine the time it took for the fluid layers of the precursors of the celestial bodies to wrap around once they reached their orbits, we need to know the length of these fluid layers. I just showed you the formula for the length of the fluid layers of the precursors of the celestial bodies. I also showed you earlier that, before the precursors began to gather into their final shape, they spent some time traveling from where they were born to the position of their orbit, where the final gathering would take place. I helped you to look at this like how you can carry a spaghetti noodle from your kitchen to your dining room before wrapping it around a fork. Now that we know the length of those fluid layers, we still need to know the speed at which they were wrapped around before we can calculate the duration of the wrapping around of these fluids to form spherical bodies. That speed (similar to the speed at which the aforementioned spaghetti noodles were wrapped around a fork) is what I called the "Nathanael-Israel Israel Gathering Speed". The next segment is just about that.

15.2.2. To know the speed at which celestial bodies got their spherical shape, don't ignore the Nathanael-Israel Israel Gathering Speed

We know that celestial bodies such as Earth, the Moon, and the Sun are spherical. But to get that shape, their precursors were curled at a certain speed. The "Nathanael-Israel Israel Gathering Speed" is the name I gave to the speed at which the precursors of the celestial bodies were wrapped around to give the celestial bodies their final shape, which, in the case of the bodies I am interested in, is spherical.

While studying the motion of the precursors of the celestial bodies using my turbulence breakthrough, I discovered that the speed I mentioned is called the orbital speed of the celestial bodies (see chapter 2). In other words, the speed at which the precursors of the celestial bodies were wrapped around is also what they use to revolve around the body they orbit today. It is like wrapping the aforementioned spaghetti noodles around a fork at a certain speed and letting the fork carrying the noodles continue the motion that initiated the wrapping. Today, people know that celestial bodies orbit other bodies, but it never crossed their mind why? Why is the Moon orbiting the Earth at a specific speed, and why is the Earth orbiting the Sun at a specific speed? In general, why are planets orbiting a primary star, and why are satellites orbiting their primary planet? I just gave you the answer, and you may not believe me (and you don't have to) until the math I do in the next chapter reveals the shocking proof for my story!

Anyway, something happened to the precursors of the celestial bodies so that they could start turning around their primary bodies. The movement that positioned the precursor into orbit was also used to wrap it around. Look at this as a gigantic fluid (e.g., water) filament moved by a current, then curled to form a kind of vortex, which continued moving on the path it was launched. In other words, using an advancement I made in understanding turbulence in the precursors of the celestial bodies, I showed that the wrapping of the precursor's fluid layer occurred at about the current orbital speed of the celestial bodies. This also implies that the movement of celestial bodies around the bodies they orbit today is a consequence of how these bodies were formed. Hence, the Moon is not turning around the Earth by chance, and, in its turn, the Earth is not turning around the Sun by chance. Did you ever ask yourself that question? Ignore what I just said, and you may continue to be one of those who think that celestial bodies just magically got their speed or shape through random events, which made things collide and amass. As you see in the next chapter, the data I reviewed contradicted that myth. Therefore, in summary, the "Nathanael-Israel Israel Gathering Speed" of the celestial bodies I am interested in can be written as:

Nathanael-Israel Israel Gathering Speed of the Earth = Earth's Orbital Speed

(Equation 18)

Nathanael-Israel Israel Gathering Speed of the Moon = Moon's Orbital Speed

(Equation 19)

Nathanael-Israel Israel Gathering Speed of the Sun = Sun's Orbital Speed

(Equation 20)

Please look at the "Nathanael-Israel Israel Gathering Speed" as the speed at which the aforementioned spaghetti noodle is wrapped around a fork. Now that I showed you my formula of the speed at which the precursor of the celestial bodies was wrapped around once it reached orbit, we finally have the final component of the formula for the duration of that wrapping. It is like knowing the length of the aforementioned spaghetti noodle and the speed at which you want to wrap it around a fork. To know how long the wrapping around would take, all you need to do is divide the length of the noodle by the speed of the wrapping around. Likewise, knowing the "Nathanael-Israel Israel Gathering Distance" and the "Nathanael-Israel Israel Gathering Speed", we can easily calculate the "Nathanael-Israel Israel Gathering Time", which is the duration required to wrap around the precursor's fluid layers once the precursor reaches orbit. In the next chapter, I will do just that. In Figure 6, I sketched how some fluid layers were collected into unique bodies.

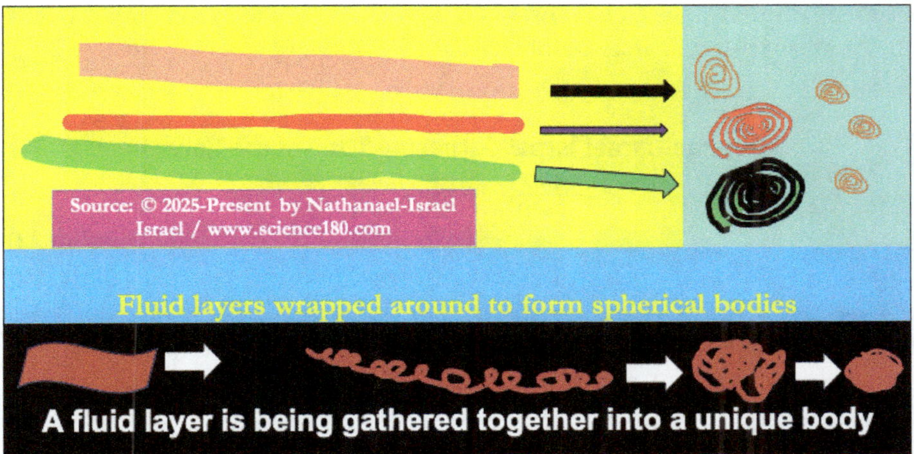

Figure 6: Picture showing fluid layers gathered together inside the precursors of celestial bodies

15.2.3. The only formula to scientifically calculate the duration of the gathering together of the precursor of the celestial bodies: Nathanael-Israel Israel Gathering Time

Because some people may not remember every step we took to get here, please allow me to remind you of the meaning of the "Nathanael-Israel Israel Gathering Time". It is the duration of time the linear fluid layers of the precursor of the celestial bodies took to swirl, or curl, to transition from being linear to becoming a spherical body. I showed you that, just before they were wrapped around, the fluid layers of the precursors were as long as the circumference of the celestial bodies they birthed. It is the circumference of these celestial bodies that I called "Nathanael-Israel Israel Gathering Distance". I also told you that they were wrapped around at a speed termed the "Nathanael-Israel Israel Gathering Distance," which equals the orbital speed of the celestial bodies as of today. Then, by dividing the length of those fluid layers by the wrapping speed, you can obtain the wrapping duration. Hence:

$$\textbf{Nathanael-Israel Israel Gathering Time} = \frac{\text{Nathanael–Israel Israel Gathering Distance}}{\text{Nathanael–Israel Israel Gathering Speed}}$$

(Equation 21)

Considering the specific formulas I developed for the "Nathanael-Israel Israel Gathering Distance" (see Equations 15-17) and the "Nathanael-Israel Israel Gathering Speed" (see Equations 18-20) of the Earth, the Moon, and the Sun, the "Nathanael-Israel Israel Gathering Time" of these bodies is:

$$\textbf{Nathanael-Israel Israel Gathering Time of the Earth} = \frac{2 \text{ x } 3.14 \text{ x Earth's Radius}}{\text{Earth's Orbital Speed}}$$

(Equation 22)

$$\textbf{Nathanael-Israel Israel Gathering Time of the Moon} = \frac{2 \text{ x } 3.14 \text{ x Moon's Radius}}{\text{Moon's Orbital Speed}}$$

(Equation 23)

$$\textbf{Nathanael-Israel Israel Gathering Time of the Sun} = \frac{2 \text{ x } 3.14 \text{ x Sun's Radius}}{\text{Sun's Orbital Speed}}$$

(Equation 24)

What I just showed you is the formula for the time it took for the precursors of the celestial bodies to collect together into spherical bodies once their fluid layers reached their orbits. In other words, if we can consider the fluid layers of the precursors of the Earth, Moon, and Sun as long spaghetti noodles, the duration of time I just showed you is how long it took for each of these gigantic spaghetti noodles to wrap around and form a special celestial body. Why is that formula important? Because we need to add that to the formula of the time it took for the precursor of the celestial bodies to move from where they were born to where they were gathered together into spherical bodies. Please note that, before being gathered together into their orbits, the precursors of the celestial bodies had a linear journey, similar to how the aforementioned spaghetti noodle traveled from the kitchen to the dining room before being wrapped around a fork. Just as we need to add the duration of the kitchen-dining room travel to the duration of the wrapping of the noodle around a fork to get the total time elapsed from the beginning of the start of the journey of the noodle from the kitchen all the way to the end of its wrapping around a fork, so also to know the duration of the formation of the celestial bodies, we need to add the "Nathanael-Israel Israel Linear Time" and the "Nathanael-Israel Israel Gathering Time". I hope you are getting me. In my book *Turbulent Origin of the Universe*, you will find the details about: what are the precursors, where did they come from, what are the wrapping around, and what is causing them to wrap around?

Before I continue, I would like to really thank you for your consideration and the time you are putting into understanding this difficult topic. But by the time you finish this book, you will know it was worth it, and the insights you gain will transform your life.

16. Want to accurately know how to calculate the duration of the formation of the celestial bodies easily? Master the Nathanael-Israel Israel Birthdate

"Truth is ever to be found in the simplicity, and not in the multiplicity and confusion of things." — Isaac Newton

An impossible-to-ignore milestone, the "Nathanael-Israel Israel Birthdate" of the celestial bodies is a term I invented to express the duration from the beginning of the formation of the Solar System until the formation of the celestial body is complete. In *Turbulent Origin of the Universe*, I showed in detail that the formation of the Solar System began at the same time as the formation of the universe.

Considering Equation 2, the Birthdate of the celestial bodies is the sum of two components:

- Nathanael-Israel Israel Linear Time, and
- Nathanael-Israel Israel Gathering Time

In some of my previous books, the "Nathanael-Israel Israel Linear Time" is what I called "Semi-Major Axis Timescale", and the "Nathanael-Israel Israel Gathering Time" is what I called the "Circumference Timescale". Now that I have explained the variables involved in each timescale, I will finally write the complete formula for the birthdates of the celestial bodies, focusing on the Earth, Moon, and Sun. Let's start with the Earth's birthdate.

16.1. The amazing formula of the Birthdate of the Earth that you will love

Using Equation 2, and considering everything we said so far, the Birthdate of the Earth is:

Nathanael-Israel Israel Birthdate of the Earth = Nathanael-Israel Israel Linear Time of the Earth + Nathanael-Israel Israel Gathering Time of the Earth

(Equation 25)

Considering Equation 9 (Nathanael-Israel Israel Linear Time of the Earth) and Equation 22 (Nathanael-Israel Israel Gathering Time of the Earth), the previous equation became:

$$\text{Nathanael-Israel Israel Birthdate of the Earth} = \frac{\text{Earth–Sun Distance}}{\text{Sun's Escape Velocity}} + \frac{2 \times 3.14 \times \text{Earth's Radius}}{\text{Earth's Orbital Speed}}$$

(Equation 26)

In the next chapter, I plug Earth's data from Table 1 into this formula to determine the exact date Earth formed. Before we get there, hold on while I also put together the final formula for the Moon's and the Sun's birthdates first.

16.2. To accurately know how long it took for the Moon to form, pay attention to the formula of the "Nathanael-Israel Israel Birthdate of the Moon"

Applying the formula of the Birthdate of the celestial (Equation 2) to the Moon, I got:

Nathanael-Israel Israel Birthdate of the Moon = Nathanael-Israel Israel Linear Time of the Moon + Nathanael-Israel Israel Gathering Time of the Moon

(Equation 27)

Considering Equation 11 (Nathanael-Israel Israel Linear Time of the Moon with respect to the Sun) and Equation 23 (Nathanael-Israel Israel Gathering Time of the Moon), it turns out that:

$$\text{Nathanael-Israel Israel Birthdate of the Moon with respect to the Sun} = \frac{\text{Earth–Sun Distance}}{\text{Sun's Escape Velocity}} + \frac{\text{Earth–Moon Distance}}{\text{Earth's Escape Velocity}} + \frac{2 \times 3.14 \times \text{Moon's Radius}}{\text{Moon's Orbital Speed}}$$

(Equation 28)

This formula has 3 components, which are respectively considered:

- the time it took for the precursor of the Earth-Moon System to journey from about the position of the precursor of the Sun all the way to about the position of the Earth (where the precursor of the Moon was born),
- the duration of the time it took for the precursor of the Moon to journey from about the orbit of the Earth to the orbit of the Moon, and
- the amount of time it took for the precursor of the Moon to swirl or curl to form the spherical Moon we know today.

With respect to the Earth, this birthday is:

Nathanael-Israel Israel Birthdate of the Moon with respect to the Earth

$$= \frac{\text{Earth–Moon Distance}}{\text{Earth's Escape Velocity}} + \frac{2 \text{ x } 3.14 \text{ x Moon's Radius}}{\text{Moon's Orbital Speed}}$$

(Equation 29)

In other words, Equation 28 signifies the time that elapsed since the beginning of the Solar System before the Moon was formed, meaning that, in that equation, the position of the Sun is the reference. In contrast, Equation 29 indicates the time elapsed since the precursor of the Moon escaped the precursor of the Earth, meaning that the Earth is the reference there. Before you get your calculator out to do some real math, let me lay out for you the formula of the Sun's birthdate quickly.

16.3. Why don't you embrace this formula of the Nathanael-Israel Israel Birthdate of the Sun so you can have a cosmological advantage before others use it to surpass you?

Considering the formula of the Nathanael-Israel Israel Birthdate of the celestial bodies (Equation 2), the date of birth of the Sun is:

Nathanael-Israel Israel Birthdate of the Sun = Nathanael-Israel Israel Linear Time of the Sun + Nathanael-Israel Israel Gathering Time of the Sun

(Equation 30)

Bearing in mind Equation 13 (i.e., Nathanael-Israel Israel Linear Time of the Sun) and Equation 24 (i.e., Nathanael-Israel Israel Gathering Time of the Sun), it turns out that the formula for the date of birth of the Sun is:

$$\text{Nathanael-Israel Israel Birthdate of the Sun} = \frac{\text{Mercury–Sun Distance}}{\text{Sun's Escape Velocity}} + \frac{2 \times 3.14 \times \text{Sun's Radius}}{\text{Sun's Orbital Speed}}$$

(Equation 31)

So far, I have shown you the thought process and the formula I developed to calculate the formation times of celestial bodies, with an emphasis on the Earth, the Moon, and the Sun. Before I say anything else, I want to remind you that I left out many unnecessary details. If I had decided to demonstrate everything in detail, I would have ended up writing another hundred-page book, just like the 700+ pages I wrote for *Turbulence Origin of the Universe*. Therefore, for those who want to build on my research, so they don't reinvent the wheel or devise hypotheses that contradict the foundations of my findings, I recommend they start with my other books. This approach will save them time and even their lives, for theorists can easily take the wrong roads, which can mislead them in their journey to test, imitate, or repeat my efforts to decode the real origin of the universe.

Now that I have explained how I developed the Turbulent Split-Gathering Framework that served as the foundation of my inquiry, we have finally come to an exciting part: plug the data calculated by top scientists throughout the ages (see Table 1) into the formula above (Equations 26, 28, and 31) to see what the story is saying. Afterward, I will scientifically test whether such a story exists elsewhere. We are getting close to our objective. Are you excited? But before that, I will recap this cornerstone chapter.

TAKE-HOME MESSAGE OF SECTION 3: DEVELOPING THE TURBULENT SPLIT-GATHERING METHODOLOGICAL FRAMEWORK

In this chapter, I shared my thoughts on how all celestial bodies had a precursor that went through a journey I divided into two: the Nathanael-Israel Israel Linear Journey and the Nathanael-Israel Israel Gathering Journey. Each of these journeys is defined by a distance and a speed, which allows us to know its duration. By adding their durations, I came up with the formula for the birthdates of the celestial bodies, which is the total time it took for them to be fully formed and put into orbit. But before I got there, I told you the "Turbulent Primary Materia", the common ancestor or precursor of all celestial bodies, went through turbulent processes, which split its bulk and gathered its daughter bodies following specific rules.

Using the term "split-gathering" that I coined, I showed you that the initial matter in the universe went through processes that shaped its daughter bodies. Just as a woman can become pregnant, and the egg will grow, pass through developmental stages, until it is born at a certain time, and then mature, passing through other developing stages before reaching an adult phase, so also, the precursors of all celestial bodies went through changes before becoming the adult celestial bodies we know today. I showed you that the precursors were, at one point, fluid layers stacked on top of one another, moving at a specific speed. I indicated that the way the precursor split-gathered explains how some bodies are organized as a primary body orbited by secondary bodies.

I told you that the precursor of the secondary bodies (meaning the bodies orbiting another body) escaped the precursor of the primary body (meaning the body it orbits), at the Nathanael-Israel Israel Linear Speed, which, for celestial bodies, is about the escape velocity of the primary body. I also showed that the Nathanael-Israel Israel Linear Distance (i.e., the distance the precursors traveled before reaching the orbit where they were collected together into spherical bodies) can be calculated by referencing the distance between the celestial bodies relative to the one they orbit. Before being wrapped around at the Nathanael-Israel Israel Gathering Speed (which for celestial bodies is about their orbital speed), the precursors of the celestial bodies were as long as the circumference (Nathanael-Israel Israel Gathering Distance) of the bodies they became.

I calculated the Nathanael-Israel Israel Linear Time by dividing the Nathanael-Israel Israel Linear Distance by the Nathanael-Israel Israel Linear Speed. Likewise, I calculated the Nathanael-Israel Israel Gathering Time by dividing the Nathanael-Israel Israel Gathering Distance by the Nathanael-Israel Israel Gathering Speed. Adding the Nathanael-Israel Israel Linear Time and Nathanael-Israel Israel Gathering Time, I obtained the formula for the Nathanael-Israel Israel Birthdate of the celestial bodies. I showed you the components of this formula, whether it was the distance the precursor traveled, the speed of travel, the length of the fluid layers before being gathered, or the speed of wrapping or swirling. Now that you know all the ingredients in this formula, let's test it and determine the actual dates of birth of the Earth, Moon, and Sun, based on the unique turbulence perspective I pioneered. Finally, I also told you that you can get the details of the math in my book *Turbulent Origin of the Universe.*

Section 4: Want to Have a Total Peace of Mind About the Exact Date of Birth of the Earth, Moon, and Sun Without Taking Sides Between Science and Faith? Together, Let Us Simply Evaluate and Apply My Turbulent Split-Gathering Methodological Framework to the Formation of the Earth, Moon, and Sun!

"Mathematics is not about numbers, equations, computations, or algorithms: it is about understanding." — William Paul Thurston

"Scientific knowledge makes our lives better when applied with concern for human welfare and environmental protection... there is no question that science and technology can produce abundance so that no one has to go without." — Jacque Fresco

"What we know is a drop, what we don't know is an ocean." — Isaac Newton

17. What is in this section for me?

After showing you the methodology I used to design the birthdate formula for the celestial bodies, I will now apply it to calculate the dates of birth for the Earth, the Moon, and the Sun. I will also use images to recount the formation of these celestial bodies in chronological order. Each time I do any math, I will ensure I break it down into simple terms that even laypeople can understand. I will be relying on the data in Table 1 and the formula I developed in the previous chapter. I encourage you to use a calculator to verify this calculation with me. After this step, I will test in the next chapter whether my findings match any story in the world. If that sounds like a good plan, let's get down to it!

If you would like to verify this math yourself, please remember to convert units properly when going from seconds to hours. For example, because there are 60 seconds in a minute and 60 minutes in an hour (meaning that one hour is 3600 seconds), to convert a duration in seconds to hours, divide the duration in seconds by 3600.

18. Can this Nathanael-Israel Israel Birthdate formula prove that the Earth was formed within days?

Considering Equation 9, and knowing the following:

- Earth-Sun distance = 149,600,000 km, and
- Escape velocity of the Sun = 617.6 km/s (see Table 1). I found that:

$$\textbf{Nathanael-Israel Israel Linear Time of the Earth} = \frac{149{,}600{,}000\ \text{km}}{617.6\ \text{km/s}} = 67.29\ \textbf{hours} = \textbf{2.804 days}$$

(Equation 32)

As a reminder, these 67.29 hours are the amount of time it took for the precursor of the Earth-Moon system to journey from about the precursor of the Sun to about the orbit of the Earth, where the precursor of the Earth was born.

Bearing in mind Equation 22, and knowing that the radius of the Earth is 6378.14 km, meaning the circumference is 40,054 km (see Equations 14 and 15), while the orbital speed of the Earth is 29.7 km/s (see Table 1), it turns out that:

$$\textbf{Nathanael-Israel Israel Gathering Time of the Earth} = \frac{2 \times 3.14 \times 6378.14\ \text{km}}{29.7\ \text{km/s}} = \frac{40{,}054\ \text{km}}{29.7\ \text{km/s}} = \textbf{22.42 minutes}$$

(Equation 33)

These 22.42 minutes are the amount of time it took for the fluid layers of the precursor of the Earth to be gathered together to form the Earth once they reached the orbit of the Earth. In other words, once the precursor of the Earth was born, it took 22.42 minutes for its fluid layers to wrap around and form the special Earth.

Considering Equation 25, Equation 32, and Equation 33, I discovered that:

Nathanael-Israel Israel's Birthdate of the Earth = 67.29 hours + 22.42 minutes = 67.66 hours = 2.82 days

(Equation 34)

If you still wonder what these 2.82 days mean, let me remind you that they are the total time it took for the Earth to form (Figure 7).

Figure 7: Formula of the Nathanael-Israel Israel Birthdate of the Earth

19. The birthdate of the Moon that no one has ever learned at Harvard University or any Ivy League institution yet

Considering Equation 11, and knowing that (see Table 1):

- Earth-Sun distance = 149,600,000 km
- Escape velocity of the Sun = 617.6 km/s
- Earth-Moon distance = 384,400 km
- Escape velocity of the Earth = 11.2 km/s, I discovered:

Nathanael-Israel Israel Linear Time of the Moon with respect to the Sun

$$= \frac{\mathbf{149{,}600{,}000\ km}}{\mathbf{617.6\ km/s}} + \frac{\mathbf{384{,}400\ km}}{\mathbf{11.2\ km/s}} = \mathbf{67.286\ hours} + \mathbf{9.54\ hours} = \mathbf{76.826\ hours = 3.2\ days}$$

(Equation 35)

Here,

- The 67.286 hours is the Nathanael-Israel Israel Linear Time of the Earth, i.e., the time that passed before the precursor of the Moon was formed. In other words, it is the amount of time it took for the precursor of the Earth-Moon System to journey from about the precursor of the Sun to the position where the Earth-Moon system split and birthed the precursor of the Moon.
- The 9.54 hours is the Nathanael-Israel Israel Linear Time of the Moon with respect to the Earth, i.e., the time the precursor of the Moon took to move from where it was born to the orbit of the Moon, where it would be collected into the spherical Moon. In other words, after the precursor of the Moon escaped the precursor of the Earth, it traveled for 9.54 hours before reaching the orbit of the Moon, where it was collected into the Moon.
- The 76.826 hours or 3.2 days is the Nathanael-Israel Israel Linear Time of the Moon with respect to the Sun, i.e., the duration or the total amount of time with reference to the beginning of the formation of the Solar System that passed before the precursor of the Moon reached the Moon's orbit, ready to be wrapped around to form the Moon. This means that from the moment the Solar System began to take shape, 76.826 hours passed before the fluid layers of the precursor of the Moon reached the Moon's orbit, ready to be collected into the spherical Moon.

Using Equation 23 and knowing that the radius of the Moon is 1738.1 kilometers (meaning that its circumference is 10,915.27 kilometers (see Equation 14 and 16)), I found that:

Nathanael-Israel Israel Gathering Time of the Moon =

$$\frac{2 \times 3.14 \times 1738.1 \text{ km}}{1.022 \text{ km/s}} = \frac{10{,}915.27 \text{ km}}{1.022 \text{ km/s}} = 2.97 \text{ hours}$$

(Equation 36)

These 2.97 hours are the amount of time required for the Moon's precursor to collect together into the spherical Moon once it arrived at the position of the Moon's orbit.

Adding the 2.97 hours to how long it took for the precursor of the Moon to move from the position of the precursor of the Earth to the Moon's orbit, I got 12.5 hours (i.e. 9.53 hours + 2.97 hours), meaning that, after the precursor of the Moon was birthed, it took 12.5 hours before it was fully formed and put on orbit. These 12.5 hours are what I call the "Nathanael-Israel Israel Birthdate" of the Moon with respect to the position of the Earth (Equation 29).

Nathanael-Israel Israel Birthday of the Moon with respect to the position of the Earth = 9.53 hours + 2.97 hours = 12.5 hours

(Equation 37)

To know the Nathanael-Israel Israel Birthdate of the Moon with respect to the beginning of the Solar System, we can proceed in two ways:

- Add the amount of time that elapsed before the precursor of the Moon was born (Equation 32) to the time it took for it to journey from where it was born until it was fully shaped and put into orbit (Equation 37): 67.29 hours + 12.5 hours = 79.79 hours, or
- Add the duration of time (with reference to the beginning of the formation of the Solar System) that passed before the precursor of the Moon reached the Moon's orbit and ready to be gathered together in a spherical body (Equation 35) to the amount of time it took for the fluid layers of the precursor of the Moon to be gathered into the spherical Moon once they arrived at the orbit of the Moon (Equation 36): 76.826 hours + 2.97 hours = 79.79 hours (Figure 8)

Therefore,

Nathanael-Israel Israel Birthdate of the Moon with respect to the Sun = 67.286 hours + 9.54 hours + 2.97 = 67.286 hours + 12.5 hours = 76.826 hours + 2.97 hours = 79.79 hours = 3.32 days

(Equation 38)

In the previous section, I explained what each number in Equation 38 means. I purposely put all of them so you can follow along more easily if you want to do this math yourself, depending on the route or option you prefer.

Figure 8: Formula of the Nathanael-Israel Israel Birthdate of the Moon

20. Proving that the Sun was formed on the 4th day is really easy if you pay attention to the Nathanael-Israel Israel Birthdate of the Sun

Before I calculate the Nathanael-Israel Israel Birthdate of the Sun using Equations 30 and 31, I will first recall a few data I will use (see Table 1):

- Radius of the Sun = 695,700 kilometers

- Circumference of the Sun = 2 x 3.14 x 695,700 kilometers = 4,368,996 kilometers (see Equations 14 and 17)
- Sun-Mercury distance = 57,909,000 km
- Escape velocity of the Sun = 617.6 km/s
- Orbital speed of the Sun: 19.4 km/s

Let me first calculate the elements in Equation 30. Using Equation 13, I found out that:

$$\textbf{Nathanael-Israel Israel Linear Time of the Sun} = \frac{57{,}909{,}000 \text{ km}}{617.6 \text{ km/s}} = \mathbf{26.046 \text{ hours}}$$

(Equation 39)

These 26.05 hours are the maximum amount of time it took for the fluid layers of all the bodies orbiting the Sun to escape the precursor of the Sun, so that the latter could be ready to wrap its fluid layers around to form the spherical Sun. Using Equation 24, I discovered that:

$$\textbf{Nathanael-Israel Israel Gathering Time of the Sun} = \frac{2 \text{ x } 3.14 \text{ x } 695{,}700 \text{ km}}{19.4 \text{ km/s}} = \frac{4{,}368{,}996 \text{ km}}{19.4 \text{ km/s}} = \mathbf{62.58 \text{ hours}}$$

(Equation 40)

These 62.56 hours represent the total amount of time the fluid layers of the Sun's precursor took to curl or wrap around to form the spherical Sun after the fluid layers of all the bodies orbiting the Sun escaped the precursor of the Sun.

Using Equation 30 in light of Equations 39 and 40, the date of birth of the Sun is 88.63 hours (i.e., 26.046 hours+ 62.56 hours). You can also reach the same date using Equation 31, which breaks down Equation 30. Therefore, the date of birth of the Sun (see Figure 9 for illustration) is:

$$\textbf{Nathanael-Israel Israel Birthdate of the Sun} = \frac{57{,}909{,}000 \text{ km}}{617.6 \text{ km/s}} + \frac{2 \text{ x } 3.14 \text{ x } 695{,}700 \text{ km}}{19.4 \text{ km/s}} = \frac{57{,}909{,}000 \text{ km}}{617.6 \text{ km/s}} + \frac{4{,}368{,}996 \text{ km}}{19.4 \text{ km/s}} = \mathbf{26.046 \text{ hours}} + \mathbf{62.58 \text{ hours}} = \mathbf{88.63 \text{ hours}} = \mathbf{3.693 \text{ days}}$$

(Equation 41)

People who attend Science180 Academy (www.Science180Academy.com) are usually shocked when I show them this mathematics that uses data collected by top scientists throughout the centuries to tell a story that supports the formation of the celestial bodies in a matter of days, not billions of days. I am also available for speaking engagements concerning any aspects of this demonstration. Learn more about that at www.Science180.com/Speaking.

Figure 9: Formula of the Nathanael-Israel Israel Birthdate of the Sun

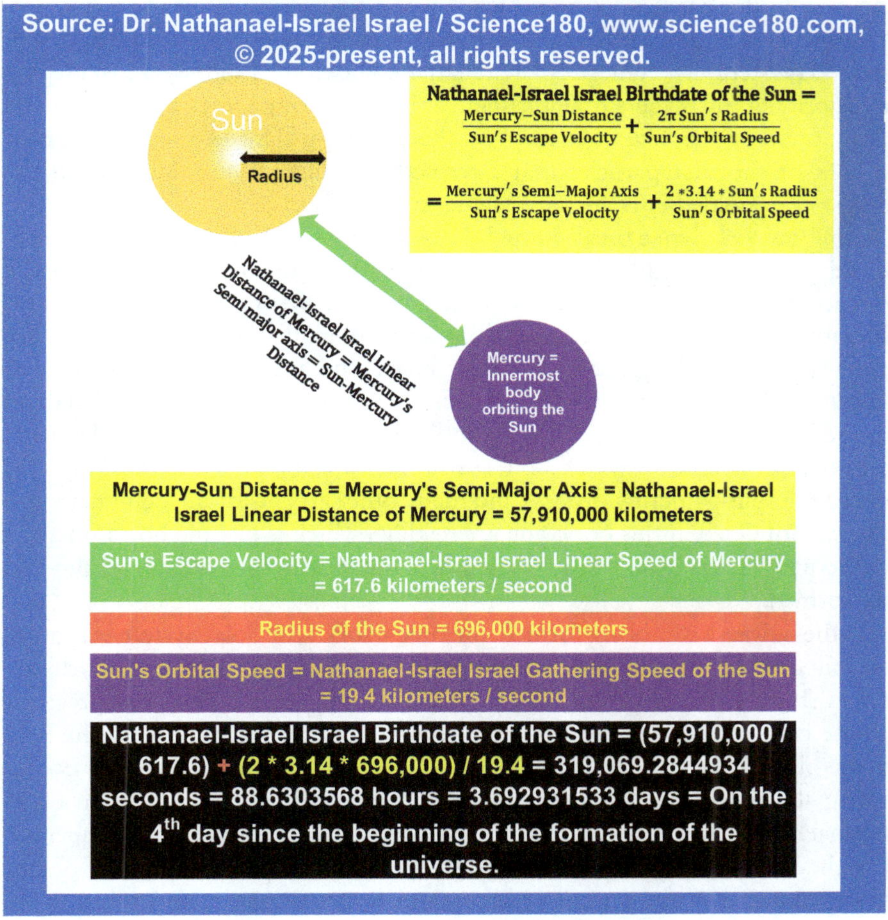

I understand that the steps we took before reaching this level are not easy for some of my readers. Won't it be great if I can recap everything I've told so far in this book in a language that refreshes your mind and leaves you with something you will never forget? Therefore, thanking you for your thoughtfulness and time, I will next drive home the main points of the story of the formation of the Earth, Moon, and the Sun using some images that may help you understand.

21. The makeover of the historic story of the formation of the Solar System that took the world by surprise!

Based on the turbulence breakthrough that I broke down in 10 books (see www.Israel120.com/books) I published before the one you are now reading, here is the story of the formation of the Solar System with an emphasis on the formation of the Earth, Moon, and Sun.

In the beginning of the formation of the universe, the bulk of the initial matter appeared mysteriously. Where did that matter come from? For the sake of helping the majority of my readers to understand at least the formation of the universe, I will not dwell on theological concepts at this point. I will instead summarize how the bulk of the first particle (that I termed Turbulent Prima Materia) in the universe was molded into the universe we know today.

Using the turbulence breakthrough I made, I showed that the initial matter in the universe was disturbed. It then underwent a series of breakups and the collection together of its daughters into various precursors. I invented the term split-gathering to describe the process by which the initial matter in the universe and its subsequent daughter bodies were broken up and gathered together into precursors of various bodies, which in turn underwent changes that molded them into the bodies and systems of bodies in the universe today. Other books that I wrote (e.g., *Turbulent Origin of the Universe, Turbulent Origin of Chemical Particles*) detailed the split-gathering process and many other aspects of my demonstration that I cannot detail in this book you are reading, which is just the bottom line of my discovery. I will now briefly explain how the Earth, the Moon, and the Sun were formed.

As the universe was being formed, the precursor of the Solar System was born, then split into the precursor of the Sun and the precursor of all the bodies orbiting the Sun (Figure 10). It was as if the precursor of the Solar System was pregnant with the precursor of the Sun and the precursor of the bodies orbiting the Sun. For the birth of these 2 daughter bodies to happen, the precursor of the bodies orbiting the Sun escaped the precursor of the Sun at about 617.6 km/s, the Nathanael-Israel Israel Linear Speed of the Earth (the escape velocity of the Sun). Once they were born, the precursor of the Sun and the precursor of the bodies orbiting the Sun were organized into fluid layers. Right after its birth, the precursor of the Sun started swirling or wrapping around at the Nathanael-Israel Israel Gathering Speed of the Sun, 19.4 km/s (i.e., the orbital speed of the Sun), to collect itself into a spherical body. But due to the length of its fluid layers, 4,368,996 km, the Nathanael-Israel Israel Gathering Distance of the Sun (which is the circumference of the Sun calculated using the Sun's radius, 695,700 km), many days would pass before the fluid layers of the precursor of the Sun could amass themselves into the gigantic Sun we know today. In the meantime, the precursor of the bodies orbiting the Sun continued its Nathanael-Israel Israel Journey.

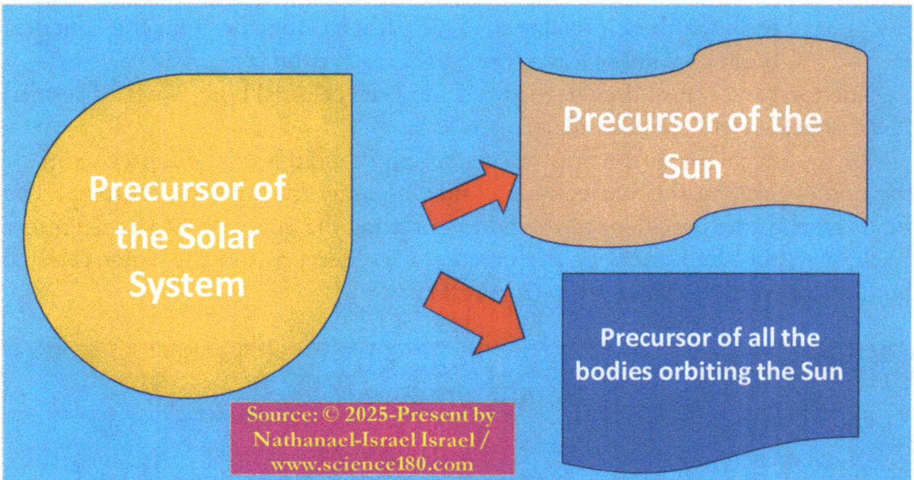

Figure 10: Precursor of the Solar System split-gathering into 2 precursors: the precursor of the Sun and the precursor of the bodies orbiting the Sun

At this stage, the fluid layers of the precursors of all the bodies orbiting the Sun today were all together, inside a single body, as a pregnant woman carrying her embryos in her belly. Just like how embryos pass through developmental stages before becoming babies, so also to a great extent the fluid layers of the precursor of the bodies orbiting the Sun went through turbulent processes before birthing the corresponding celestial bodies we know today. The same was true for the precursor of the Sun itself.

The journey of the fluid layers of the precursors from their birth all the way to their complete gathering together into their corresponding bodies was dominated by turbulence; hence, I labelled this journey the "Nathanael-Israel Israel Turbulent Journey". For each precursor of a celestial body, this journey was divided into two events (i.e., the "Nathanael-Israel Israel Linear Journey" and the "Nathanael-Israel Israel Gathering Journey") lasting different amounts of time:

- "Nathanael-Israel Israel Linear Time": the amount of time it took for the precursors to move from where they were born to where they were collected together and put into orbit, and

- "Nathanael-Israel Israel Gathering Time": the duration of time it took for these precursors to be gathered together into the spherical bodies we know today, once they reach orbit.

The total amount of time needed to form each celestial body is what I termed "Nathanael-Israel Israel Birthdate."

As the precursor of the Sun was gathering its fluid layers, the precursors of the bodies orbiting the Sun continued moving away from it after escaping. The fluid layers of the precursor to the bodies orbiting the Sun were organized into a stack, with the fluid layers of the precursor to Mercury and the other celestial bodies closest to the Sun on top (Figure 11).

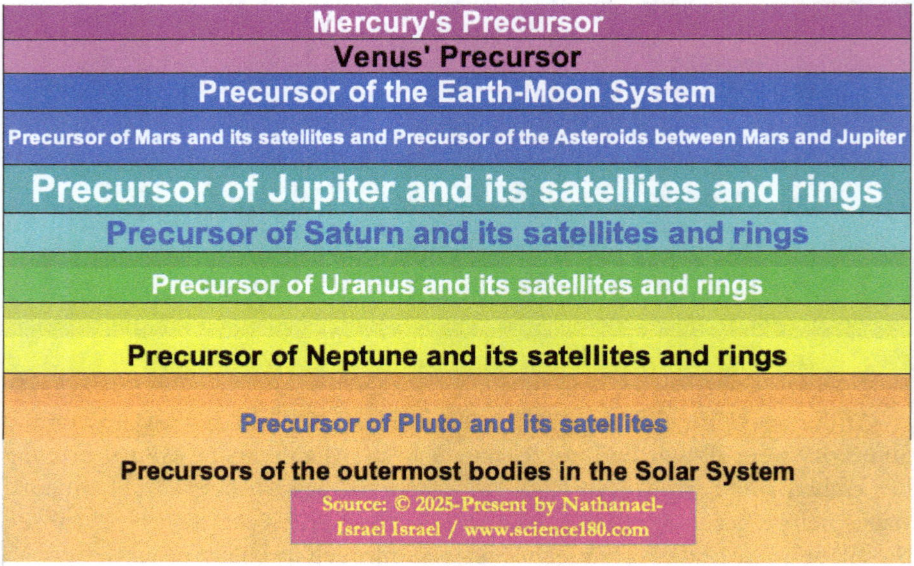

Figure 11: Position of the fluid layers of the precursor of Mercury, the precursor of the bodies located between Mercury and the Earth, the precursor of the Earth-Moon system, and the precursor of the bodies (in the Solar System) located beyond the Earth-Moon system. After the precursor of the bodies orbiting the Sun escaped the precursor of the Sun, its fluids were organized in layers. The layers of Mercury's precursor were on top. I neglected the small asteroids between Mercury and the Sun for simplicity purposes.

The fluid layers of the outermost bodies in the Solar System were at the bottom of the stack. In contrast, the fluid layers that would become the precursor to the Earth-Moon system were embedded within fluid layers beneath the fluid layers of all the celestial bodies between the innermost bodies of the Sun and the Earth. In other words, the fluid layers of all the bodies located between the Earth and the Sun were located above the fluid layers of the precursor of the Earth-Moon System. And beneath the fluid layers of the precursor of the Earth-Moon System were the fluid layers of all the bodies located beyond the Earth-Moon system in the Solar System. Put another way, during the formation of the Solar System, the fluid layers of all the bodies beyond the Earth-Moon System were located beneath the fluid layers of the precursor to the Earth-Moon System.

During the journey of the remainder of the fluid layers of the precursor of the bodies orbiting the Sun, fluid layers of the precursor of celestial bodies split from the main stack according to their position, with the top layers splitting first and the bottom layers splitting last. Once the precursors of the bodies orbiting the Sun escaped the precursor of the Sun, they started a very long journey. Imagine this as a stack of fluid layers travelling together at a certain speed. Because fluids are not solid, these layers moved at different speeds, some layers sliding past others and affecting surrounding layers. For instance, precursors of the planetary systems formed as fluid layers atop one another. Because the layers at the bottom carried more weight, they moved more slowly and gave rise to planets that orbit the Sun more slowly than those born from the top layers. Although I am trying to put this in simple language, you must understand that, here, I overlook many scientific details, so that everybody (including nonscientists) can get something out of this historic discovery that took the world by surprise in 2025 after I, a Beninese-American scientist, sacrificed 12 years of my life to incubate it in my mind.

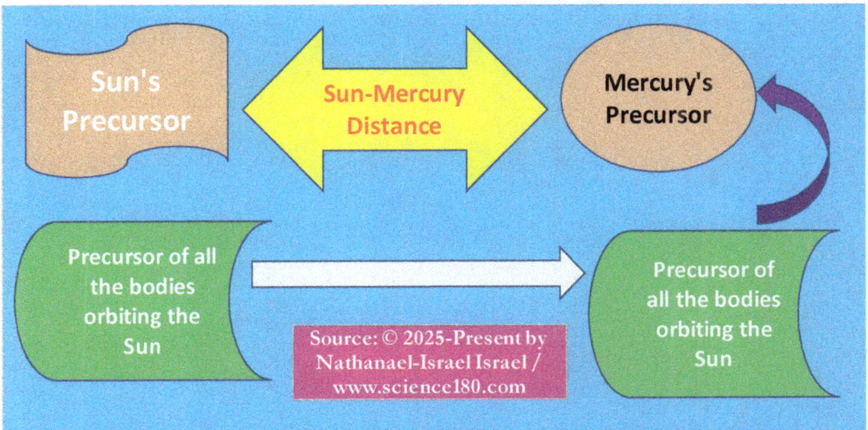

Figure 12: The precursor of all the bodies orbiting the Sun traveled the "Nathanael-Israel Israel Linear Distance" of Mercury, meaning the Sun-Mercury distance before birthing Mercury's precursor

Travelling at 617.6 km/s (the escape velocity of the Sun, which scientists calculated since the days of Sir Isaac Newton, meaning more than 400 years ago), for about 26.046 hours after they escaped the precursor of the Sun, the fluid layers of the precursor of the bodies orbiting the Sun found themselves at about the Nathanael-Israel Israel Linear Distance of Mercury, 57,909,000 km away from the precursor of the Sun. Therefore, the fluid layers of the precursor of Mercury (Figure 12) escaped the stack and were collected into Mercury. Hence, that distance is the actual distance separating the Sun and Mercury. By the time the 26.046 hours passed, all the fluid of the precursor of the bodies orbiting the Sun left the Sun, and the precursor of the Sun was fully formed and ready to start collecting itself into a spherical body. Hence, I estimated that the fluid layers of the precursor of the Sun began to collect together into the spherical Sun about the time the precursor of the closest body orbiting the Sun split from the precursor of the bodies orbiting the Sun. In other words, the fluid layers of the Sun's precursor started wrapping around to collect into the spherical Sun about 26.046 hours since the beginning of the formation of the Solar System. These 26.046 hours are the Nathanael-Israel Israel Linear Time of Mercury, which is about the Nathanael-Israel Israel Linear Time of the Sun.

Once the fluid layers of the precursor of Mercury (or whichever celestial body will be found as the closest to the Sun) escaped the stack of fluid layers of the precursor of the bodies orbiting the Sun, the latter continued its journey away from the precursor of the Sun. The precursor of the Earth-Moon System was still embedded somewhere between fluids (Figure 13) and waited for its turn before splitting from the stack.

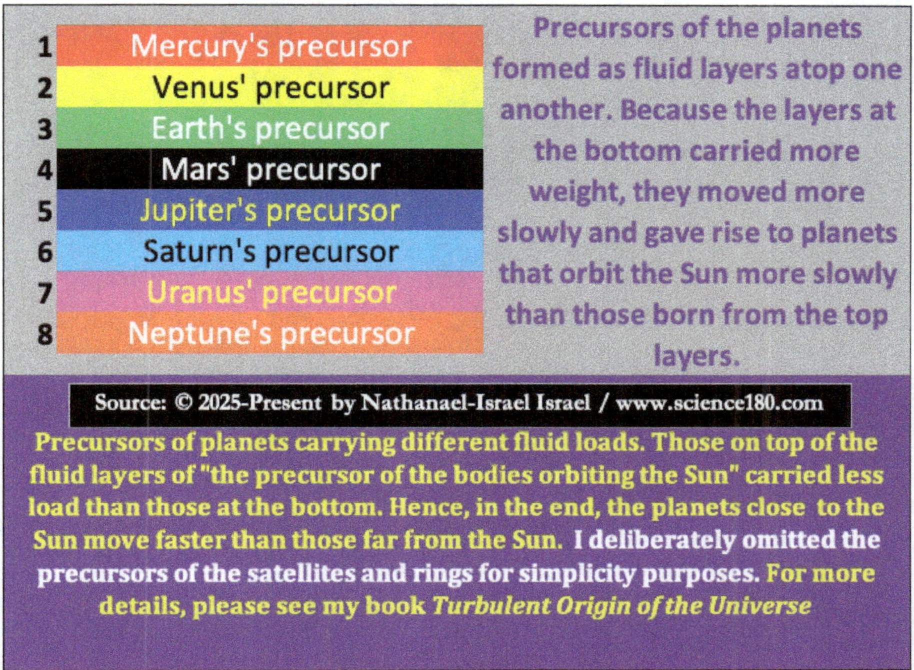

Figure 13: Why planets have different speed and some move faster than others

Journeying at 617.6 km/s (i.e., the Nathanael-Israel Israel Linear Speed of the Earth, also known as the escape velocity of the Sun), for 67.29 hours (meaning 2.804 days, which is the Nathanael-Israel Israel Linear Time of the Earth), the rest of the stack of the fluid layers of the bodies orbiting the Sun were 149,600,000 km (which I termed the Nathanael-Israel Israel Linear Distance of the Earth, also known as the Earth-Sun distance) away from the precursor of the Sun. There, the fluid layers of the precursor to the Earth-Moon system lay atop the remaining fluid layers of the bodies orbiting the Sun (Figure 14). It split from all the fluid layers beneath it to form the precursor to the Earth-Moon system (Figure 15), which, in turn, quickly split into the precursors to Earth and the Moon. In my book *Turbulent Origin of the Universe*, I detailed some high-level mathematics that helped me understand that this split was very violent and happened within a few minutes.

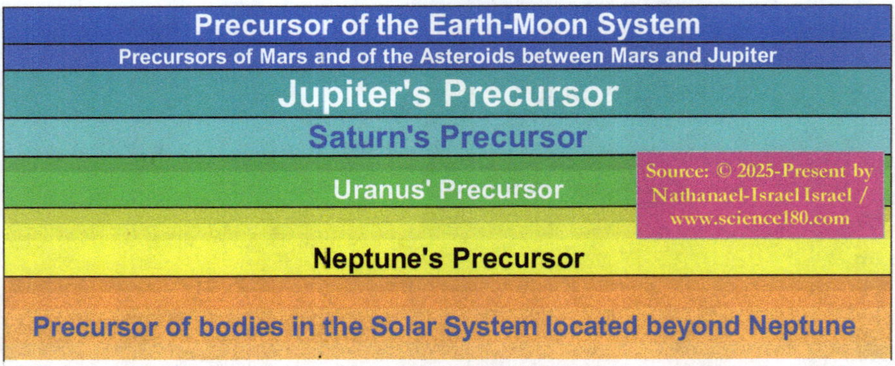

Figure 14: The precursor of the Earth-Moon system about the split from the remainder of the fluid layers of the precursor of the bodies orbiting the Solar System, after all the fluid layers of the precursor of the bodies located between the Earth and the Sun split from the stack of fluid layers of the precursor of the bodies orbiting the Sun. In this graph, I deliberately omitted the precursors of the satellites and rings for simplicity purposes. You can find details about them in my book "Turbulent Origin of the Universe"

In other words, about the Nathanael-Israel Israel Linear Time of the Earth, 67.29 hours, after the formation of the Solar System began, the precursor of the Earth and the precursor of the Moon formed, having escaped the remaining fluid layers of the precursor of the Earth-Moon System. Before elaborating on what happened to the precursor of the Earth and the precursor of the Moon once they were formed, let me quickly explain the fate of the remainder of the fluid layers of the bodies orbiting the Sun.

In fact, after the precursor of the Earth-Moon System split from the rest of the stack of the fluid layers of the precursor of the bodies orbiting the Sun, the latter was lighter, and continued its journey, and the fluid layers split from the rest according to their position, forming the asteroids and planetary systems located beyond the Earth in the Solar System. The size, the speed, the position, and many other characteristics of these fluid layers explained the turbulence that occurred in them and which led to the formation of specific bodies, including all the planets, satellites, and asteroids in the Solar System, which, so this book does not become another big one, I chose to ignore in this book. For those who want to know more, please refer to *Turbulent Origin of the Universe*. In that book, you will also see what happened in the other stellar systems, galaxies, and their clusters in the universe. Because I already explained all of those things, you don't need to waste your time trying to figure them out or reinvent the wheel, which is copyrighted by someone else! And when it comes to the explanation of the turbulent origin of the universe through the Turbulent Split-Gathering Framework that I broke down here and in my other books, I am honored to be the discoverer.

The precursor of the Earth-Moon System was born after the remainder of the precursor of the bodies orbiting the Sun traveled the Nathanael-Israel Israel Linear Distance of the Earth (Earth-Sun distance). See Figure 15.

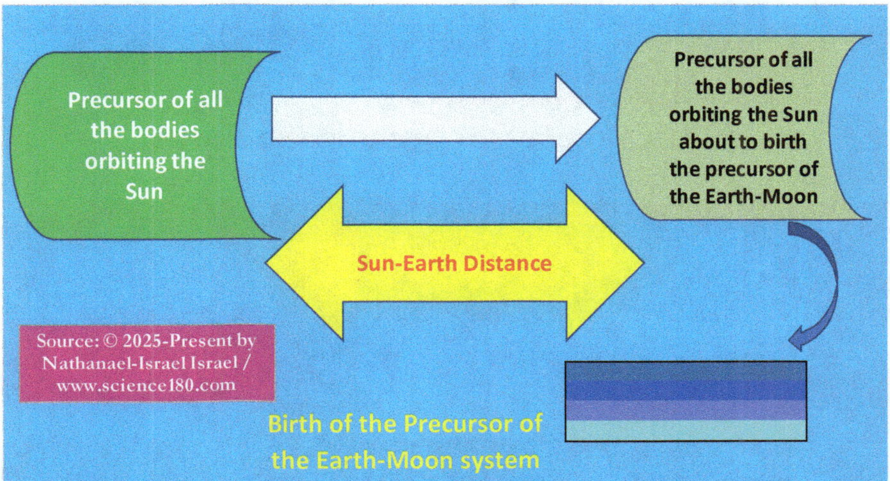

Figure 15: Precursor of all the bodies orbiting the Sun traveled about Nathanael-Israel Israel Linear Distance of the Earth, meaning the Sun-Earth distance, 149,600,000 km, before birthing the precursor of the Earth-Moon system, which then split-gathered into the Earth's precursor and the Moon's precursor

Then, the precursor of the Earth-Moon split into the precursors of Earth and Moon, both of which were organized into fluid layers. For the precursor of the Earth-Moon system to birth its daughter bodies, the precursor of the Moon escaped the precursor of the Earth at about 11.2 km/s (i.e., Nathanael-Israel Israel Linear Speed of the Moon, which is the escape velocity of the Earth). Afterward, the precursor of the Moon journeyed for about 384,400 km (i.e., Nathanael-Israel Israel Linear Distance of the Moon, known as the Earth-Moon distance) before arriving at the position where its fluid layers were ready to be collected into the spherical Moon. Before I explain what happened to the precursor on the Moon once it arrived in the Moon's orbit, let me finish talking about the formation of the Earth first.

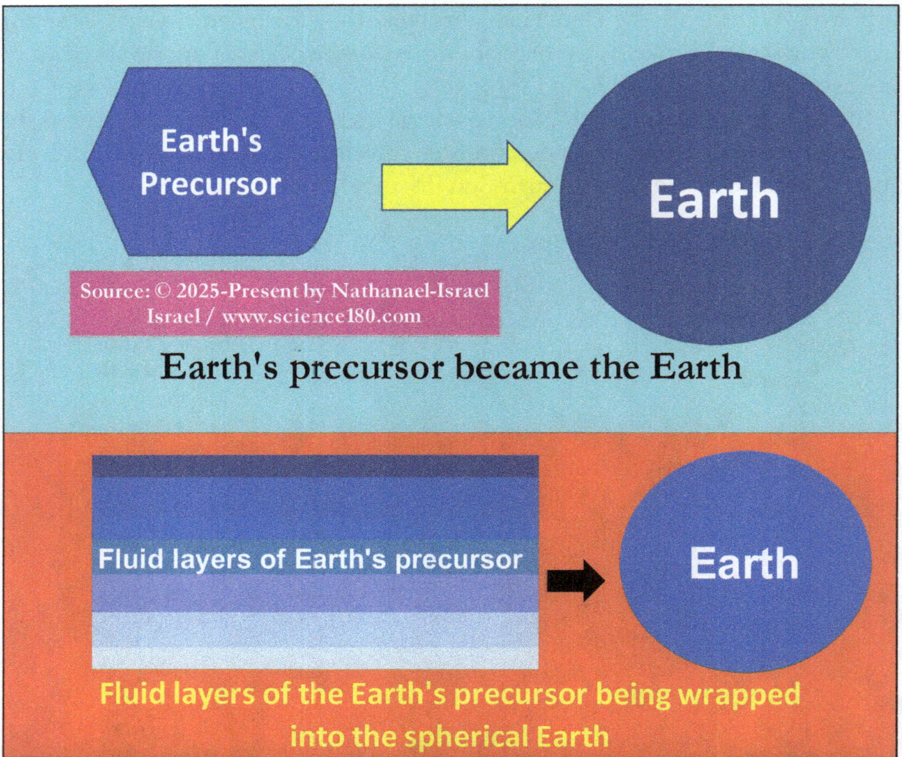

Figure 16: Earth's precursor became the Earth

After the fluid layers of the precursor of the Moon split and began moving away from the precursor of the Earth, the fluid layers of the precursor of the Earth began to swirl and gather together into the Earth. Long as 40,054 km (which is the Nathanael-Israel Israel Gathering Distance of the Earth, calculated as the circumference of the Earth), the fluid layers of the Earth's precursor were wrapped around at about 29.7 km/s (i.e., Nathanael-Israel Israel Gathering Speed of the Earth = orbital speed of the Earth). Within 22.42 minutes (i.e., Nathanael-Israel Israel Gathering Time of the Earth), all the fluid layers of the Earth's precursor were gathered together into the spherical Earth, whose radius is 6378.14 km. In other words, after the precursor of the Earth-Moon system birthed the Earth's precursor, it took just 22.42 minutes for all the fluid layers of the precursor of the Earth to be collected into the Earth we know today. Since, from the start of the Solar System, about 67.29 hours (i.e., Nathanael-Israel Israel Linear Time of the Earth) passed before the Earth's precursor was formed, and since it took 22.42 minutes for it to be gathered into the Earth, the total amount of time that elapsed before the Earth was fully formed was: 67.29 hours + 22.42 minutes = 67.66 hours. Knowing that 1 day equals 24 hours, these 67.66 hours are equivalent to 2.82 literal days. In short, the Earth formed on the 3rd day, more specifically, 2.82 days after the formation of the Solar System began. These 2.82 days are what I call the Nathanael-Israel Israel's Birthdate of the Earth (Figures 16-18).

Figure 17: Fluid layers of Earth's precursor were rolled like spaghetti noodles around a fork

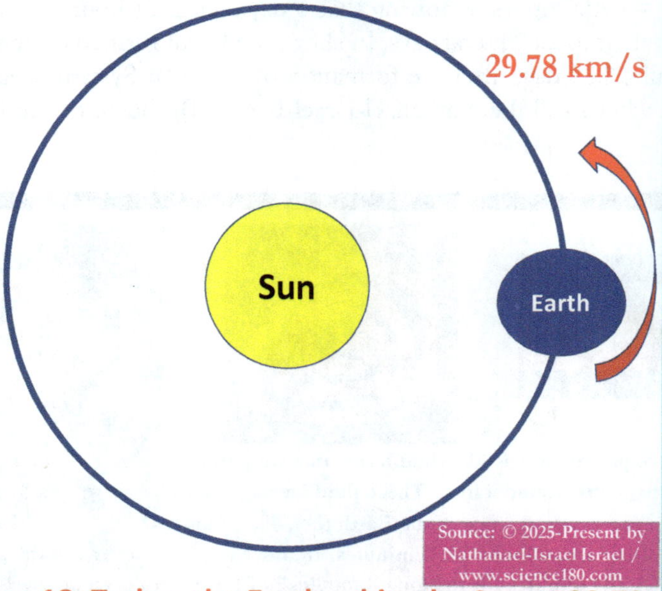

Figure 18: Today, the Earth orbits the Sun at 29.78 km per second, which is the Nathanael-Israel Israel Gathering Speed of the Earth, because the Earth's precursor was wrapped around at about that speed

Up until this point, the precursor of the Sun was still gathering its fluid layers to form the Sun. Moreover, by the time the Earth formed, the precursor of the Moon was still on a journey toward the Moon's orbit. In fact, because it was travelling at 11.2 km/s (i.e. = Nathanael-Israel Israel Linear Speed of the Moon = Earth's escape velocity), it took 9.54 hours (i.e. Nathanael-Israel Israel Linear Time of the Moon with respect to the Earth) for the precursor of the Moon to travel the 384,400 km (i.e. Earth-Moon distance, which I termed the Nathanael-Israel Israel Linear Distance of the Moon with respect to the Earth) to move from the precursor of the Earth to the orbit of the Moon, where it would be collected into the spherical Moon (Figure 19). In other words, after the precursor of the Moon escaped the precursor of the Earth, it traveled for 9.54 hours before reaching a point where it was collected into the Moon. Because, with respect to the beginning of the Solar System, it took 67.286 hours before the precursor of the Moon was formed, in total, with reference to the beginning of the formation of the Solar System, 76.826 hours (i.e. 67.286 hours + 9.54 hours) passed before the precursor of the Moon reached the position (Moon's orbit), where it was ready to be wrapped around to form the Moon. These 76.826 hours is what I called the "Nathanael-Israel Israel Linear Time of the Moon with respect to the Sun", the duration of time it took for the initial particles that would become the Moon to journey from about the position of the precursor of the Sun all the way to the orbit of the Moon, where these particles were finally organized into fluid layers ready to be wrapped around to form the Moon.

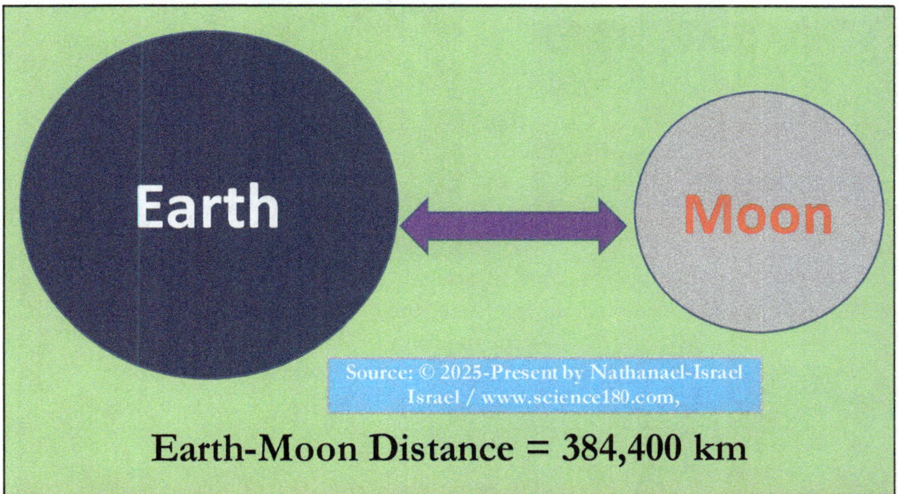

Figure 19: Nathanael-Israel Israel Linear Distance of Moon (with respect to the Earth)

About the time they arrived on the orbit of the Moon, the fluid layers of the precursor of the Moon were 10,915.27 km long. That is what I called the "Nathanael-Israel Israel Gathering Distance of the Moon", also known as the circumference of the Moon, calculated using the Moon's radius (1738.1 kilometers). The spiraling, swirling, or curling of these fluid layers of the Moon's precursor happened at 1.022 km/s, the "Nathanael-Israel Israel Gathering Speed of the Moon", also known as the Moon's orbital speed. Therefore, it took 2.97 hours for these 10,915.27 km long fluid layers of the Moon's precursor to wrap around at 1.022 km/s to form the Moon once they arrived in the Moon's orbit. These 2.97 hours represent the "Nathanael-Israel Israel Gathering Time of the Moon" (Figure 20). Because, after escaping the precursor of the Earth, the precursor of the Moon traveled for 9.54 hours (i.e., the Nathanael-Israel Israel Linear Time of the Moon with respect to Earth, also known as the Moon-Earth distance) before the beginning of its curling around as a whole, the total amount of time elapsed with respect to the birth of the Moon's precursor before the Moon was formed was 12.5 hours (i.e., 9.54 hours + 2.97 hours). These 12.5

hours are what I termed the "Nathanael-Israel Israel Birthday of the Moon with respect to the Earth". In other words, after the precursor of the Moon was formed, it took 12.5 hours before the Moon itself was fully formed and put into orbit.

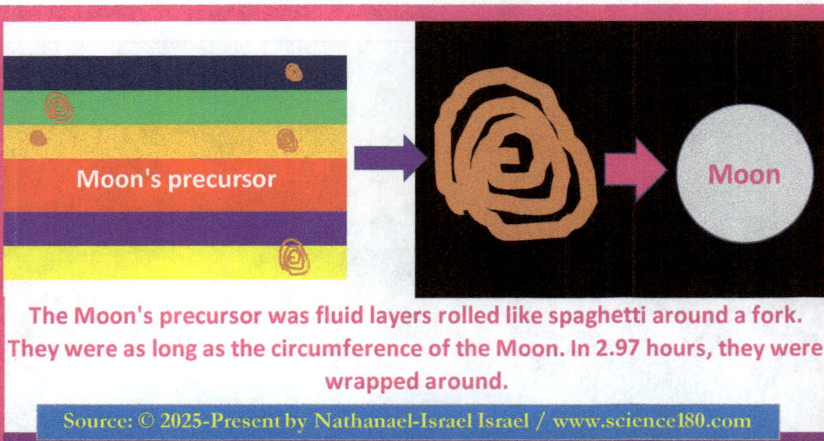

Figure 20: Nathanael-Israel Israel Gathering Time of the Moon, meaning how long the fluid layers of the Moon's precursor took to roll around to form the spherical Moon

Since the beginning of the formation of the Solar System, about 67.29 hours passed before the precursor of the Moon was born; the formation of the Moon was fully completed 79.79 hours (i.e., 67.29 hours + 12.5 hours) after the beginning of the formation of the Solar System. Considering the Earth's date of birth, the formation of the Moon was completed about 12.5 hours after the Earth was fully formed (Figure 21).

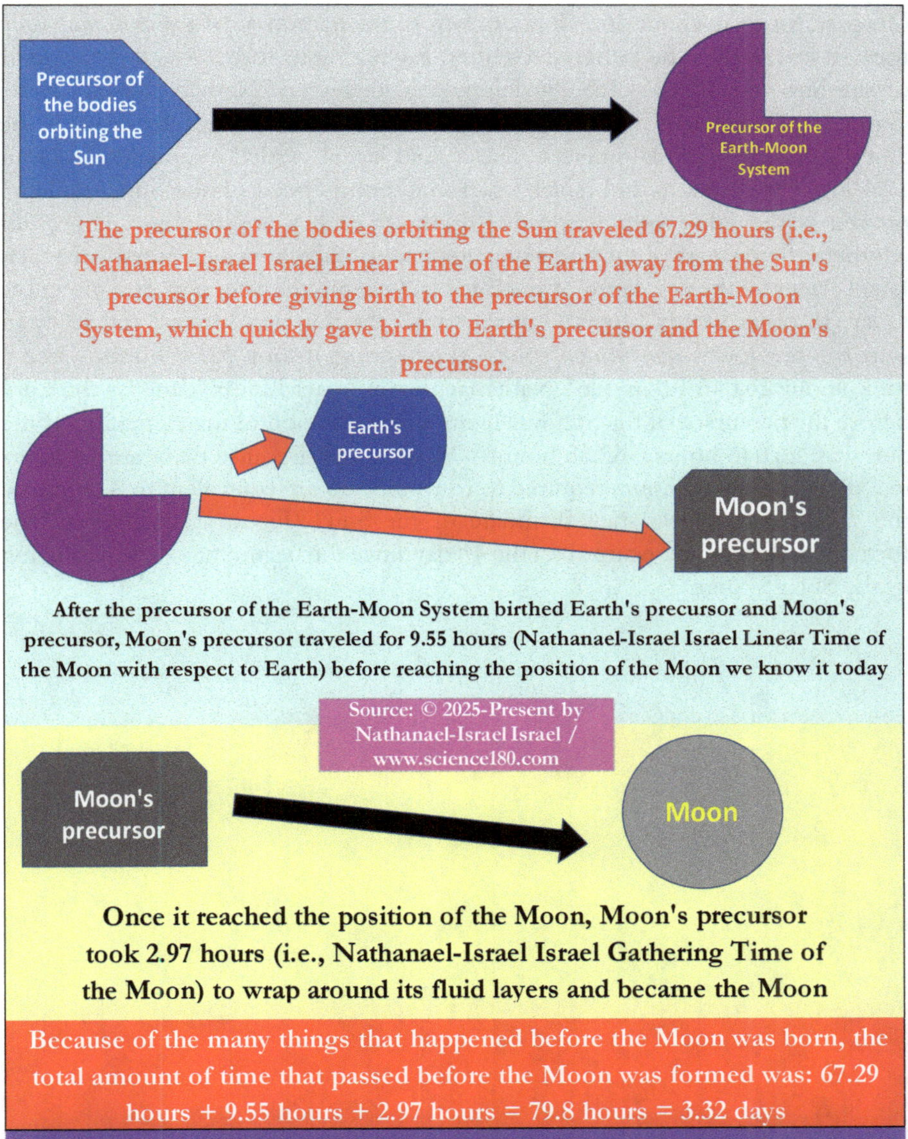

Figure 21: Nathanael-Israel Israel Birthdate of the Moon: The time it took for the Moon to be formed

Incredibly, until this point, the Sun had not yet fully formed. Why? Because the fluid layers of the Sun's precursor were very large and were curled at a relatively slow speed, it took a lot of time for them to be gathered together into a spherical body. Considering the "Nathanael-Israel Israel Linear Time of

Mercury," it took about 26.046 hours before the precursor of the Sun was fully formed and ready to be gathered together. Right after its formation, the precursor of the Sun was about 4,368,996 kilometers long (i.e., "Nathanael-Israel Israel Gathering Distance of the Sun"), calculated as the Sun's circumference using the Sun's radius, 695,700 kilometers. These fluid layers swirled or spiraled at about 19.4 km/s (i.e., Nathanael-Israel Israel Gathering Speed of the Sun," which I proved is the Sun's orbital speed. Curling at such a speed, these 4,368,996-kilometer fluid layers of the Sun's precursor took 62.58 hours (i.e., the Nathanael-Israel Israel Gathering Time of the Sun) to become the spherical Sun we know today. In other words, after the precursor of the Sun was formed and ready to start the previously mentioned spherical gathering, it took 62.58 hours. Since it took about 26.046 hours (i.e., Nathanael-Israel Israel Linear Time of the Sun) before the precursor of the Sun was formed, the duration of the formation of the Sun was: 26.046 hours+ 62.58 hours = 88.63 hours. Because there are 24 hours in a day, these 88.63 hours required to form the Sun are equivalent to 3.693 days: the "Nathanael-Israel Israel Birthdate of the Sun" (Figure 22). In short, the formation of the Sun occurred on the 4th day since the beginning of the formation of the Solar System.

Figure 22: Nathanael-Israel Israel Birthdate of the Sun, i.e., how long it took for the Sun to be formed

Knowing the date of birth of the Earth (67.66 hours) and of the Moon (79.79 hours), the Moon was therefore formed about 12.13 hours after the Earth, and the Sun was formed 8.84 hours after the Moon, meaning that the Sun was formed about 20.97 hours after the Earth. Who would have imagined that the gigantic Sun was formed after the Earth and the Moon? It took so long before the fluid layers of the Sun's precursor were collected in the Sun because they were more than 100 times longer than the fluid layers of the Earth's precursor and were curled at a speed lower than that of the fluid layers of the Earth's precursor. So, while the precursor of the Moon traveled a long distance before being curled into the Moon at a speed lower than that of the Sun's precursor, the Moon was still formed before the Sun because the layers of the Sun's precursor were too long to be collected together. The Moon was formed after the Earth because its precursor had to travel for many more hours after escaping the Earth's precursor and being collected into the Moon. These mysteries that no scientist has ever proved before me explain why the Earth was formed before the Moon, and while the Sun was formed after the Earth and its only satellite.

The details of this mathematical demonstration and its relation to the formation of other celestial bodies (including planets, satellites, stars, galaxies, and their clusters) in the universe can be found in my other books, particularly *Turbulent Origin of the Universe* and others. Figure 23 summarizes some key elements involved in calculation of the Nathanael-Israel Israel Birthdate of the Earth, the Moon, and the Sun.

Although I did not start my research with a hypothesis, to comply with the scientific requirement, here are a few hypotheses I could have stated, considering my focus on the formation of the Earth, the Moon, and the Sun:

- The formation of the Earth was completed on the 3rd day after the formation of the Solar System began
- The formation of the Moon was completed on the 4th day after the beginning of the formation of the Solar System began
- The formation of the Sun was completed on the 4th day after the beginning of the formation of the Solar System

As the scientific evidence showed, in light of my theory, the scientific data support these hypotheses, proving that the Earth's formation was really completed on the 3rd day, while the formation of the Moon and the Sun was on the 4th day, with respect to the beginning of the formation of the Solar System (Figure 24).

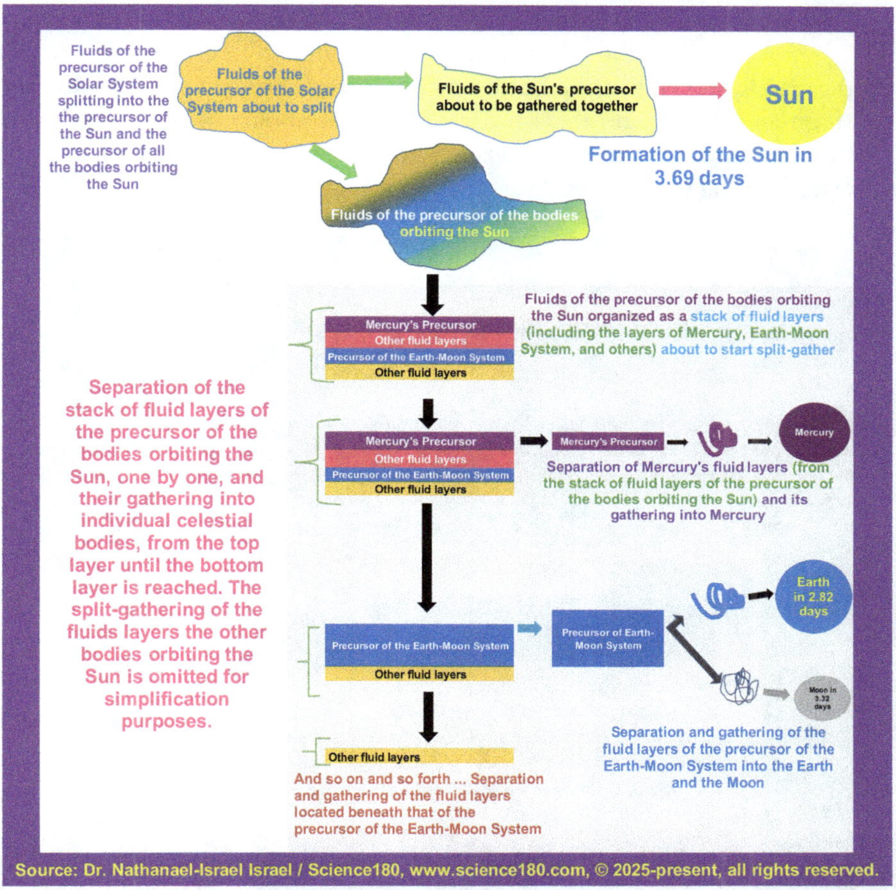

Figure 23: Split-gathering of the fluid layers of the precursor of the Solar System into many precursors including that of the Sun, Mercury, the Earth and the Moon

TAKE-HOME MESSAGE OF SECTION 4: EXACT DATE OF BIRTH OF THE EARTH, THE MOON, AND THE SUN

In this chapter, I used the machinery or mechanism I discovered for the formation of the Earth, Moon, and Sun, according to the perspective of the turbulence breakthrough I spearheaded, to calculate the exact amount of time it took to form these celestial bodies as follows:

- Nathanael-Israel Israel's Birthdate of the Earth = 67.66 hours = 2.82 days
- Nathanael-Israel Israel Birthdate of the Moon = 79.79 hours = 3.32 days
- Nathanael-Israel Israel Birthdate of the Sun = 88.63 hours = 3.693 days

Therefore, using the Turbulent Split-Gathering Framework that I spearheaded, I proved that the Earth was completely formed on the 3rd day, and the Moon and the Sun on the 4th day after the beginning of the Solar System (Figure 24). I also did a historic makeover of the story of the formation of the universe, focusing on the Solar System!

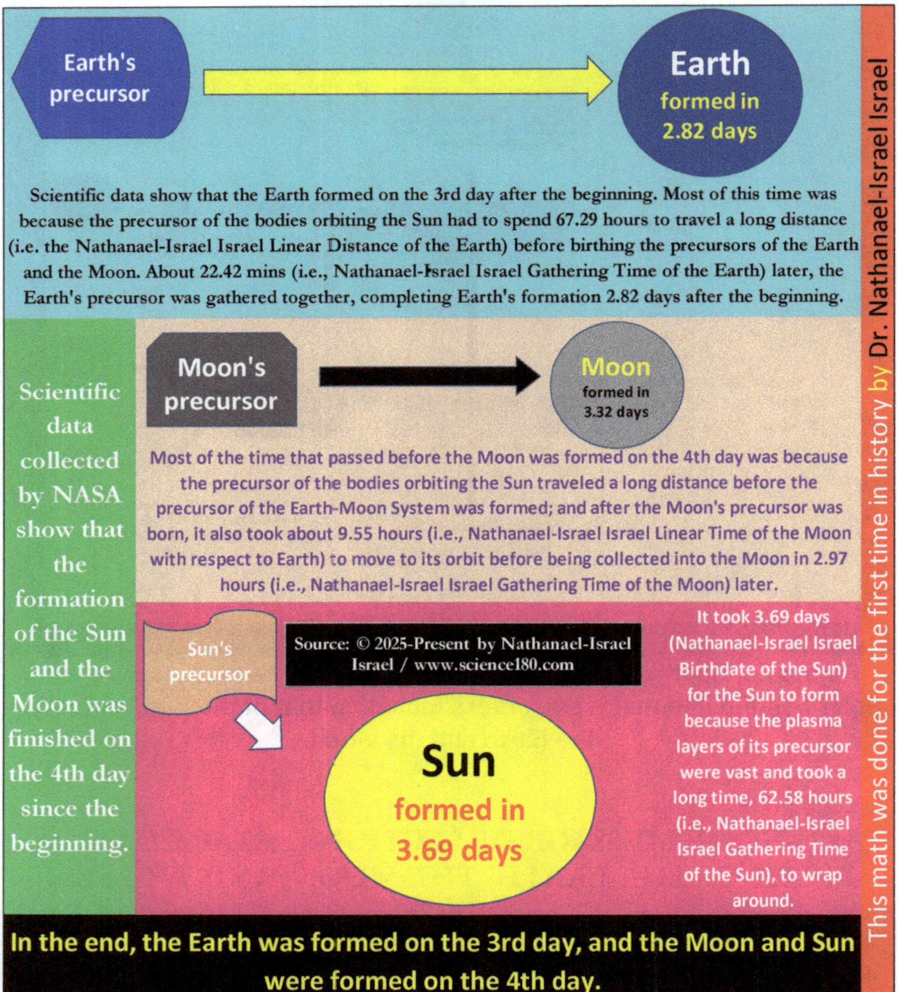

Figure 24: What the scientific data said about the birthdate of the Earth, the Moon, and the Sun.

Section 5: Evaluating and Verifying my Turbulent Split-Gathering Theory and Troubleshooting Failed Theories

"Science is a beautiful gift to humanity; we should not distort it." — A. P. J. Abdul Kalam

"When dealing with people, remember you are not dealing with creatures of logic, but creatures of emotion." — Dale Carnegie

"To improve is to change; to be perfect is to change often." — Winston Churchill

"You will find science not antagonistic, but helpful to Religion," — Lord Kelvin (William Thomson)

"We should be careful to get out of an experience only the wisdom that is in it and stop there lest we be like the cat that sits down on a hot stove lid. She will never sit down on a hot stove lid again and that is well but also she will never sit down on a cold one anymore." — Mark Twain

22. To properly test if any existing framework (scientific or religious) partially or fully applies my Turbulent Split-Gathering Framework, just start like this! It is that simple

"An expert is a person who has made all the mistakes that can be made in a very narrow field." — Niels Bohr

"I'm not an answering machine, I'm a questioning machine. If we have all the answers, how come we're in such a mess?" — Douglas Cardinal

"You cannot teach a man anything; you can only help him discover it in himself." — Galileo Galilei

Does my methodological framework link to any existing theory or dogma? Do my findings agree with any universe's origin story in the scientific literature or in the world's religions?

To answer that question, I investigated cosmological theories and creation stories across the world's religions. As you can see, this theory differs from all other scientific theories ever proposed to explain the formation of the universe, whether they are scientific, philosophical, or religious. It would be very easy for me to point my fingers at and criticize everyone one by one, singling out where I know they were wrong. From the secular viewpoints to the extremist creationists, passing by the religiously non-affiliated, without forgetting all variants of freethinkers, there are reasons to challenge and call everyone out. But what would I gain from such an attack on people, some of whom have really tried to discover what I found, but who, unfortunately, made consequential mistakes they must bear, some of which are deadly?

Without diluting the implications of my discovery, I may use some diplomacy and borrow some wisdom from the renowned Dale Carnegie, who, in his famous bestseller book, *How to Win Friends & Influence People*, advises: "*Show respect for other people's opinions. Never say, "You're wrong.*" As I try to use some of this advice (well! do I have to?) so I can make some friends and convince them (don't tell them I say that, for nobody wants to be proven a loser in an argument), please consider another piece of advice that the same Dale Carnegie gave: "*If you're wrong, admit it.*"

As we compare the scientific theory I advanced in this book with the creation stories of the world, if it turns out that your beloved scientific theory, religion, or ideology is proven wrong, I hope you will admit it. I hope you will act, unlike most people who don't want to be proven wrong and who won't admit they are wrong. Otherwise, what must I, a fearless, outspoken person, do to help you then?

SECTION 5: EVALUATING AND VERIFYING MY THEORY AND TROUBLESHOOTING FAILED THEORIES

A long time ago, I may not have thought twice before throwing the truth at your face as it is, but, as I grow wiser (I hope you will agree that I am trying to be wise and have you as my friend), I understand that there are other ways to convey the truth, still without sugarcoating it with lies.

Before I share the verification and the implications of my discovery, please know that my goal is not to accuse you or anyone, but to help and equip you. For when I consider the implications of the choices regarding the acceptance or rejection of the theory I presented in this book and others I wrote (see www.Israel120.com/books), I understand that if those (believers or unbelievers) who deny it really knew what was coming next, they would not have denied its literalism. Similarly, when I consider the nonsensical and mythological arguments advanced by some who believe in the story's literal interpretation. I also realized that most of them don't even know what they are talking about. In all humility, what I have discovered has humbled me and taught me that most people ignore most of what they are talking about, even what they think they know for SURE. Therefore, no matter what I will say in the remainder of this chapter, please know that my goal is to help you, not to hurt or to prove that you have willingly done something wrong. I think that most people, even the unbelievers, want to answer some of the most asked questions I asked in this book. But the problem is that they are using the wrong methodological framework and data to "convince" themselves of the perspective they already hold. Therefore, no matter the proof presented to some people, they may not change their minds. But that should not be an excuse for not telling people the truth. Therefore, as I am about to take the risk of sharing courageously my opinions about why no one else, including you, has discovered what I have, please be open to what I want to tell you. Once again, remember! I want to be your friend. My goal is not to prove you wrong or teach you a lesson, but to share the verification of my theory and what I think is my best explanation of what went wrong. And if I can be proven wrong, I also want to be put right. Therefore, please examine my thoughts with me.

To get there, I will first review what I think went wrong with most of the world's religions, the secular world, the creationists themselves (whether Old Earth creationists, Young Earth creationists, Intelligent Design proponents), the freethinkers, without forgetting the proponents of evolution, the Big Bang theory, and all other scientific theories that attempted to address the origin questions, then I will conclude the book. Let me start with the world's religions.

23. Is it easy to test scientifically whether any religion in the world is correct? Pay attention to the code behind the Nathanael-Israel Israel Birthdate of celestial bodies

"Mathematics is the language with which God has written the universe." — Galileo Galilei

"I believe the more I study science, the more I believe in God… All my discoveries have been made in answer to prayer." — Isaac Newton

"God has been pleased to work in his material creation by laws. … The Creator governs his material works by definite laws resulting from the forces impressed on matter… The natural works of God can never by any possibility come into contradiction with the higher things that belong to our future existence." — Michael Faraday

"Science is the process of thinking God's thoughts after Him. … I wanted to become a theologian... Now, behold, God is praised by my work even in astronomy." — Johannes Kepler

Up to this point, I know some people may still doubt my discovery. What if a story told thousands of years ago could perfectly match what I discovered? How can we know if any or which of the world's religions is scientifically correct? We will answer those questions by comparing scientific facts with religious creation stories. You will know for sure if any religion is correct. No scientist has ever proven what you are about to read now. So, buckle up and let's get down to it!

I have spent a lot of time investigating creation stories across the globe (See my book *Science180 Accurate Scientific Proof of God.* Here, I will summarize just a few points sufficient to answer the main question we are interested in.

As of 2025, about 30% of the world's population is Christian, making Christianity the world's largest religion. Jews, to whom the Biblical account of creation was given, make up approximately 0.2% of the world population (the majority of them being in the US and Israel, with many increasingly returning to Israel). In contrast, about 25% of the world's population is Muslim. Therefore, more than 55% of the world's population believes in creationism. About 10% of the world is said to be "convinced atheists," and including agnostics and other unaffiliated, especially in China, most of the religiously unaffiliated worldwide. In general, about 85% of the world's population is said to identify with a religious group. Without a scientific demonstration, how can anyone judge and declare the truth teller or the liar among the world's religions, some of which both claim to believe in or deny God?

SECTION 5: EVALUATING AND VERIFYING MY THEORY AND TROUBLESHOOTING FAILED THEORIES

I discovered that all the creation stories don't agree with one another, and some religions don't even have such a story. Some religions even discourage their adepts from engaging in debates about the origin of the universe, for they don't deem such debates important. Interestingly, the Abrahamic believers (i.e., Jews, Christians, and Muslims) are the ones who believe in the creation of the universe by God. In other words, among the world's religions, only the Abrahamic religions have a creation story of a few days. But none of these stories is as chronological and detailed as the Biblical account of creation.

The Biblical account of creation was recounted by Moses about 3500 years ago, meaning about 1500 years before Jesus Christ, while the Islamic account of creation was written less than 1500 years ago (considering that Mohammed, the founder of Islam, was born around 570 and died in 632, meaning about 600 years after Jesus Christ). This means that the Biblical account of creation was told almost 2000 years before the Islamic narrative of creation.

Although the Koran and the Bible state that God created the universe in 6 days, and many people believe in that story literally, countless others reject it. However, regardless of how Abrahamic believers interpret the creation narrative in their religious texts, it is important to underscore the significant differences between the Islamic narrative of creation and the Biblical account. These differences lie in the details of the events corresponding to each day of creation:

- While the Biblical account of creation states that the formation of the Earth was completed on the 3rd day of creation, the Koran says that the Earth was formed on Sunday (1st day of creation) and on Monday (2nd day of creation).
- While the Biblical account of creation clearly says that the Moon and the Sun were formed on the 4th day of creation, the Koran says that the Moon and the Sun were formed on Friday (the 6th day of creation).

If you can recall the scientific evidence I showed earlier in this book, the formation of the Earth was completed on the 3rd day, while that of the Moon and the Sun was on the 4th day. This shows that only the Biblical account of creation matches the scientific theory I demonstrated in this book and others I wrote. For instance, in my book *Science180 Accurate Scientific Proof of God, From Science to Bible's Conclusions*, and *Reconciling Science and Creation Accurately*, I detailed how, of all the religions in the world, only Judeo-Christianity has a creation story that properly matches the scientific evidence. Figure 25 summarizes what the Bible actually says about the birthdate of the Earth, the Moon, and the Sun (Book of Genesis 1:14-19):

"In the beginning God created the heaven and the earth. 2 **And the earth was without form, and void**; and darkness was upon the face of the deep. And the Spirit of God moved upon the face of the waters. 3 And God said, Let there be light: and there was light. 4 And God saw the light, that it was good: and God divided the light from the darkness. 5 And God called the light Day, and the darkness he called Night. And the evening and the morning were the first day. 6 And God said, Let there be a firmament in the midst of the waters, and **let it divide the waters from the waters**. 7 And God made the firmament, **and divided the waters** which were under the firmament **from the waters** which were above the firmament: and it was so. 8 And God called the firmament Heaven. And the evening and the morning were the second day. 9 And God said, **Let the waters under the heaven be gathered together unto one place, and let the dry land appear: and it was so. 10 And God called the dry land Earth; and the gathering together of the waters called the Seas**: and God saw that it was good. 11 … 12 … and God saw that it was good. 13 **And the evening and the morning were the third day**. 14 And God said, Let there be lights in the firmament of the heaven to divide the day from the night; and let them be for signs, and for seasons, and for days, and years: 15 And let them be for lights in the firmament of the heaven to give light upon the earth: and it was so. 16 And **God made two great lights; the greater light [Sun] to rule the day, and the lesser light [Moon] to rule the night**: he made the stars also. 17 And God set them in the firmament of the heaven to give light upon the earth, 18 And to rule over the day and over the night, and to divide the light from the darkness: and God saw that it was good. 19 **And the evening and the morning were the fourth day**". (King James Version).

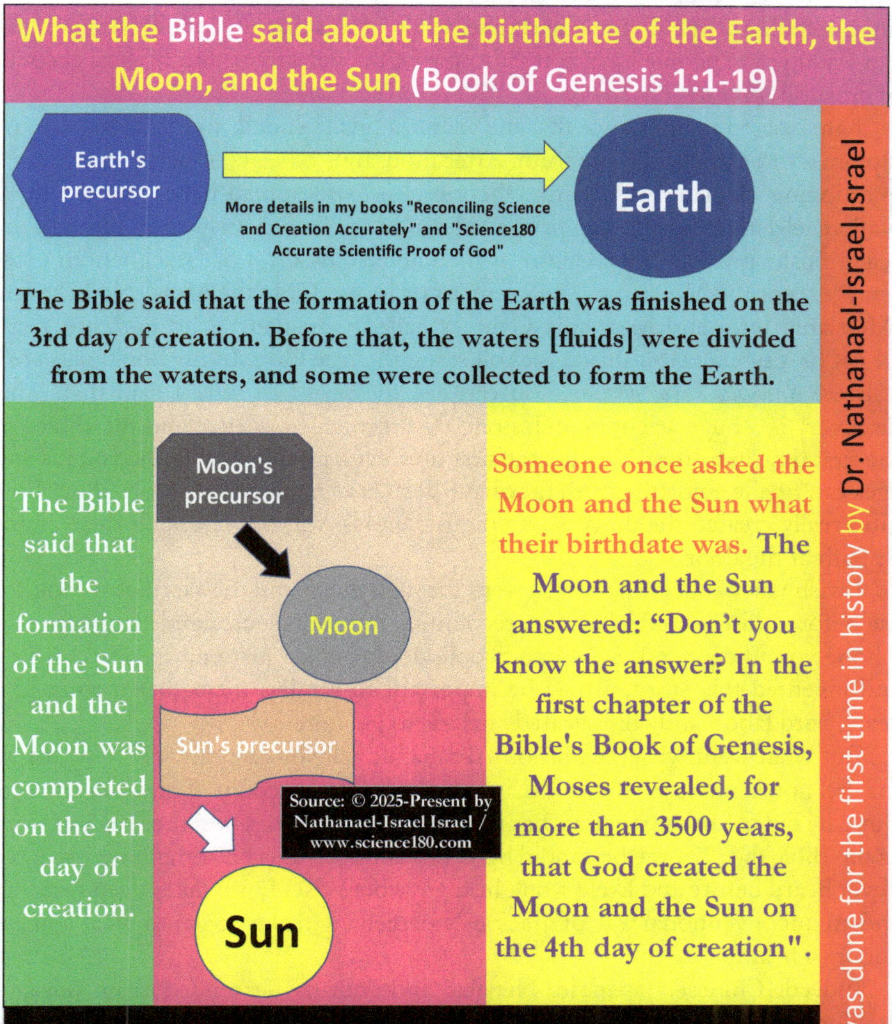

Figure 25: The Bible said in the Book of Genesis that the Earth was completely formed on the 3rd day, while the Moon and Sun were formed on the 4th day. This is exactly what Science confirmed based on the research of Dr. Nathanael-Israel Israel, a respected Beninese-American Scientist. Learn more at www.Israel120.com

But because there are many denominations or branches among the Jews and the Christians, I need to say which one has properly interpreted the creation story. To say that all Biblical believers are correct on their viewpoints about creation would be very dangerous and heretical. Some of them completely missed the interpretation of the Genesis narrative, while others got almost everything right. For 3500 years, people have sought to prove the veracity of the Biblical account of creation until my discovery.

I am excited to announce that the wait has finally ended, and I cannot use any diplomacy to camouflage the truth. After thousands of years of persistent rational questioning, demands, criticisms, theories, and widespread belief and doubt, I, Nathanael-Israel Israel, am privileged to have finally answered the call by scientifically proving the creation of the universe by God of Israel, whom most Jews (except the Messianic Jews) deny as of today. Only the Messianic believers (Messianic Jews and Messianic Gentiles called Christians) who embrace Jesus Christ or Yeshua HaMashiach are following the true God. In my book *Science180 Accurate Scientific Proof of God*, you can see the proof of why I said that. This discovery is a huge scientific milestone that reconciled science and the Bible for the first time in history. Yes, for the first time ever, people of all backgrounds and ages can finally enjoy the truth they have been craving to demonstrate about how God really created the universe, as the Bible says, without checking their brains or faith at the door.

I wish you could know what I went through in life and the sacrifice it took for me before I discovered what I am recounting to you in such simple language that I hope you understand. So, some people don't think I just got up one morning and invented this story; it may be important to remind them that it took 12+ years, hard labor, and complicated analysis to get here.

But what went wrong with non-Abrahamic religions worldwide? Why, of all the creationist stories in the world, is only the Biblical account of creation correct? But why did the other religious books in the world fail to reveal the creation story as the Bible did? Let me be candid here: I'm about to tell you something you have never heard before and likely won't hear anywhere else! For it takes deep research and an unconventional way of graciously perceiving reality to get to what you are about to hear.

Indeed, Chinese, Japanese, Nepalis, and Indians descended from lineages existing before Moses was born. Unlike the Jews and Arabs, to my knowledge, Abraham is not their ancestor, and they were probably unaware of the ancient or sacred story of creation, which the Abrahamic believers share. Hence, the eastern religions (Buddhism, Hinduism, Confucianism, etc.) born in those Asian nations have no clue about the Biblical creation story or even a corrupted version of it, like a certain Abrahamic religion (that I will address very soon) does. Most of these eastern religions did not come close to the Biblical account of creation, which is the only story that is backed by the scientific theory I presented in this book. Buddhists don't even talk about creation. Furthermore, Confucianism (named after its founder, the Chinese philosopher Confucius (551–479 BC)), the dominant philosophy, political ideology, religion, or tradition in China, did not provide any detailed account of creation. Confucianists believe that the world is going through a continuous ordering or configuration.

I think the ancestors and followers of most of these Eastern religions were frank enough not to invent a creation story and to force the Buddhists or the Confucianists to embrace it. Who knows if it is not the thirst for the truth (they ignore) that causes Buddhists, Confucianists, and others to put so much emphasis on meditation and energy balance in life …?

In contrast, Hindus have many versions of creation and many beliefs, but they are all completely different from the Genesis narrative in the Bible. Animists in Africa, the Caribbean, and other parts of the world also have many creation stories, but they are all absurd, though there are a few elements of "truth", as if their ancestors had come into contact with some creation stories. Therefore, I realized that the creation stories of the world's religions boil down to what their founders or ancestors knew or did not know about creation from Moses' narrative! Because I have already answered this question in my books title *Science180 Accurate Scientific Proof of God* and *Reconciling Science and Creation Accurately*, I will not repeat myself here. For more details, please consult those books.

Those who participate in Science180 Academy (www.Science180Academy.com) always tell me that my materials really help them unlock more details about what makes my discovery so special and why its long-overdue worldwide availability is a win for humankind seeking to understand their root. Can I give you my opinion on the big question most people ask me about the drift between Islam and Judeo-Christian religions? Why are Islam and traditional Judaism (meaning the Judaism that denies Jesus Christ as the Messiah) wrong religions? I reported some of my findings in my books:

- *Science180 Accurate Scientific Proof of God*
- *Reconciling Science and Creation Accurately*

In the next chapter, I will summarize my main points concerning Islam and the various creationist viewpoints.

24. Can we use mathematics to scrutinize the Koran and prove that Islam is scientifically incorrect? Can anybody really dare to say that?

"As a blind man has no idea of colors, so have we no idea of the manner by which the all-wise God perceives and understands all things." … "No sciences are better attested than the religion of the Bible." … "In the absence of any other proof, the thumb alone would convince me of God's existence," — Isaac Newton

"I have been blamed by men of science, both in this country and in England, for quoting the Bible in confirmation of the doctrines of physical geography. The Bible, they say, was not written for scientific purposes, and is therefore of no authority in matters of science. I beg pardon! The Bible is authority for everything it touches." — *Matthew Fontaine Maury*

Although just as Christians and Jews, Muslims believe in God (which Muslims call Allah, while the Jews call God Adonai, Yahweh, Jehovah, I am, … and the Christians refer to God as Jesus) creating the universe in 6 days (whether literal days or not), the Islamic creation narrative significantly differs from the Biblical narrative. The most striking differences between the Islamic narrative of creation and the Biblical account of creation can be found in the details of what the Koran and the Bible describe as occurring on each of the 6 days of creation.

As I introduced it in the previous segment, according to the Koran, the Earth was created on Sunday (1st day of creation) and on Monday (2nd day of creation), while the Sun and the Moon were created on Friday (6th day of creation). Although Jews and Arabs share a common ancestor, the way the creation story may have been transmitted orally throughout generations may have corrupted the Islamic account of creation. The ancestors of the Arabs may have learned part of their creation story from Abraham or his descendants, including Ismael, a direct forefather of the Arabs.

Considering the animosity between the Jews and the Arabs, the founder of Islam may have been misled by a demon who portrayed himself as the Angel Gabriel, as mentioned in the Bible as a good messenger. Either that demon misled Mohammed and/or his followers (who wrote the Koran) invented the Islamic story of creation using what they knew from the Jewish story of creation, which was known for almost 2000 years before the birth of Islam. If not, how can the Koran put the formation of the Moon and the Sun on the same day (6th day), while the Bible puts them also on the same day (but on the 4th day, which corroborates the scientific evidence advanced in my books)? I felt like someone tampered with the creation story in the Koran; otherwise, the Muslims and the Jews could still be enemies, but at least still have the same creation story. Islam for Today (2017) pointed out that the Koran doesn't contain a complete chronology of creation, and Muslim scholars do not believe in the literal meaning of the days of creation. Because science proved that the Biblical account of creation is true, I inferred that Mohammed might have not encountered a holy angel from the God of Israel. For God cannot contradict Himself. In *Science180 Accurate Scientific Proof of God*, I explained how Muslims view the Biblical account of creation as a corrupted version of God's message.

Considering the scientific evidence in my theory and the scientific timelines (see the previous chapter of this book), which match the Biblical account of creation, I strongly believe that the Koran is NOT a revelation from God that the Muslims call Allah. The math and science that I showed in my books overwhelmingly proved that the Koran is a corrupted message, which the Muslims ignore. For the same holy angel Gabriel cannot reveal Jesus Christ to Virgin Mary as the Messiah of the whole world, and then 600 years later reveal the same Jesus to Mohammed just as a simple prophet, not even the Savior. Being a God of order, the God of Israel, who shared the Biblical creation story with Moses, could not have shared a contradictory story of creation with Mohammed. The Koran just tried to paint the Jews as the enemies of God. But are all Jews children of God, as they say? No, of course. And I already addressed that issue earlier, and more details can be found in my book *Science180 Accurate Scientific Proof of God.*

But, without referring to history, being gracious, thoughtful, courageous, impartial, unconventional, caring, and really putting ourselves in the shoes of some Arabs, can we blame all Muslims for their religious "choices"—(knowing that some Muslims did not really choose to be Muslim, for they were born in some Arab nations or home, and embraced the religion of their land as most people would do until they understand the truth and be set free)—and actually win them to the God of Israel that most Israelites (even until today) deny just as those who ruled Jerusalem about 2000 years ago did to the point that they killed their own Messiah (Yeshua HaMashiach) they ignored, while looking for peace outside of Him who said (in the New Testament) that He is the Prince of Peace? Please, pay attention to the fact that I, Nathanael-Israel Israel, must tell you now!

After carefully examining the issue surrounding the discrepancy between the Biblical and Islamic narrative beyond the creation narratives and scientific proof that I spearheaded, I think that some Muslims are just practicing what their ancestors learned from their parents and so on, until now, without faith in the God of Israel (Israel that most Muslims perceive as the #1 enemy of Allah, whom they call God), because the Muslims think they can't embrace the God whom the Messianic Jews and Messianic Gentiles (called Christians) accept, but which the non-Messianic Jews (the majority of the Jews) DENY for now. Similarly, non-Messianic Jews are practicing a religion inherited from their ancestors, who made some mistakes in the interpretations of God's intent in the Bible, but who refuse to correct themselves, even when countless Messianic Jews came to embrace Jesus Christ as the Messiah. Additionally, the behaviors of some Israelites (particularly the Jews that refute Jesus or Yeshua HaMashiach as the Messiah promised in the Hebrew Bible—which some Christians called OLD Testament (as if it is obsolete)—, while holding onto the NEW Testament (that the non-Messianic Jews say is false and heretical), while some Christians support EVERYTHING the nation of Israel is doing)… made it very difficult for some Arab Muslims, whose ancestors carried and transferred onto their descendants the hurt and hate they felt toward all Jews whom, regardless of the warning of

Apostle Paul (a Pharisee that the New Testament said met Jesus and converted into Messianic Judaism, which birthed Christianity), who told the Jews that not all Jews are children of God or children of Abraham, even considering how Ishmael (the ancestor of the Arabs) was dismissed by Abraham (at the profit of Isaac, a forerunner of the Jews after Abraham, the common forefather of both Jews and Arabs), who, according to the Bible, said that the God of Israel spoke to him to dismiss Ishmael and his mother, Hagar, by sending them away into the wilderness! Wow! Well! If seen through the glance of physical proofs, ignoring the spiritual, how can the descendants of Ishmael ever believe that what the Bible says is TRUE? Hence, among other things, for thousands of years, the Arabs (descendants of Ishmael, the son that the Bible says that Abraham dismissed righteously) looked at the Biblical story as a baloney and the Jews as enemies of God, while the Jews themselves who love physics (and won more Nobel Prizes in physics and other fields than any other nation), while ignoring the spiritual, denying God they claim to like, to the point of naively killing the Son of God, therefore God, whom rose from the dead 3 days later (having hidden some of His [God's] plans to the Devil and his children, some of whom Jesus said are the Pharisees), otherwise, they would not have killed Jesus, whom needed to die anyway, so the gentiles whose ancestors hated God (at a time when some ancestors of the Jews loved God) can start loving God until the time both the nation of Israel and even many Arab nations (including Iran described as Persia in the Bible) will wake up and turn to God, whom will gladly call them my people! Therefore, why would you expect all Arabs and all Jews to naturally tell the same story about the creation of the universe? Why would you expect them to have a lasting peace deal between themselves today, while they ignore the Prince of Peace (Jesus, Yeshua HaMashiach), that the unbelieving Jews (i.e. Pharisees) killed 2000 years ago because, among other things, they said that Jesus (that most Christians and Messianic believers believe is God) said that He is the Son of God, thus God, something they considered as a blasphemy, while Islam says that Jesus (that Muslims refer to as Issa) is just a simple prophet, lower than Mohammed (that Muslims consider as the greatest prophet of Allah, even above Moses who revealed the scientific proof of God in the Genesis account of creation)! Therefore, the Pharisees (who claimed to be children of Abraham) killed their God on the cross, while saying that they were believing in and serving God, whom [Jesus] told them that they were children of Satan, the father of lies. At the same time, the Egyptians until today do not believe that the Jews ever lived in Egypt for about 430 years before God of Israel delivered them from their bondage in Egypt (Africa) while sinking the Egyptian army in the Red Sea, before the "stiff-necked" Israelites (that's how God called some of them in the Bible) rebelled against God in the Wilderness for 40 years, to the point that, before entering Canaan, God killed all of those who left Egypt, including Moses himself, except two, without forgetting that the Israelites never really possessed the Promised

Land, the Land filled with Milk and Honey! By the way, is it by chance that the Africans (i.e., Egyptians) put the Jews into slavery for 400 years and then the Westerners (including the Jews) put the Africans into slavery for 400 years in the America— transatlantic slave trade? And now, without understanding impartially the spiritual root of all of this, you want the Arab Muslims and their allies to just have a peace deal with the believers of the God of Israel? Think about this for a moment! That is the kind of thing that can happen when we neglect the spiritual roots of reality, while focusing too much on the correlation between physical things we can "see."

But now, thanks to science and my scientific breakthrough, we know who is right. Hard to say, but that is the truth that even science can verify. Yes, modern science confirms that the Biblical account of creation is true and that the Islamic creation story is false. But why that? I just told you what I think is the root cause of the Jewish and Arab fighting for thousands of years, and that war is now transferred to the Christians and the Muslims. But the Messianic Jews, the Messianic Christians, and the Arab Christians, and Arab Messianic who understand the spiritual mechanism beyond what I just said, know that the real enemy of Israel is not the Arabs, vice versa, nor the real solution to their conflict to come from any human being or country, no matter how powerful, right, and determined they will be. The real enemy of all of us is spiritual, and it is not limited to any nation, even if some nations promote evil things, while others claim to be God's nations as if God (that most people, even countless of those who confess to believe in Him, hate) died for a nation or as if salvation is national, not personal, or as if our value or ranking in the spiritual world is really based on our material or physical things, some of which we don't even merit, deserve, or properly work for!

But can we really say that any nation, political party, or organization on this Earth is really obeying God 100%, and that there is a country or party that is God's, and others that God really hate 100% and that have no one God cares for? Well, that's what some people who claim to have the truth want those they think lack the truth to embrace! It is not surprising that those biased doctrines can't solve anything sustainably! Only Jesus Christ, whom all unbelievers deny, is the solution, and in the New Testament, Apostle Paul presented Him as the one whom God the Father used to create the universe. You can learn more about that in *Science180 Accurate Scientific Proof of God* and *Reconciling Science and Creation Accurately*.

My friend, this is a very complicated puzzle that you can never decode if you don't humbly, patiently, impartially, unconventionally, and courageously seek the truth of what has been happening beyond physical restrictions. And if some Jews and Christians can team together to express their hatred toward the Muslims, vice versa,—without considering that some Muslims and non-Messianic Jews (let's say all unbelievers) are just trapped in a snare they cannot get out of without we candidly showing them love and care in how we present strategically the Messianic truth (which can be summarized into the need of humankind to believe in Jesus Christ as Lord of Savior), or without being prejudiced in how we confront evil, even within our own party, organization, or nation,—how can we otherwise help God (not just the God of Israel, but the God of all nations, the God of the whole universe, who is called in the Bible the God of Israel) to save the lost? And when I talk like this, I raise many enemies against me. But do I care about that and even about what they will try to do to me? You may not like what I am saying here, but this is the application of my holistic Turbulent Split-Gathering Framework (which cracked the code of the universe's origin) to the current state of the world's affairs regarding God and his people, whom we think are physical but spiritual.

Have you asked yourself what my name, Nathanael-Israel Israel, means? This is my real name. Among other things, I love God. I love Israel. I love many other nations (including the United States of America, where I live) as well, — God bless America—and I must tell the Truth to the whole world impartially. For I have a special place in God's heart, and He trusts me for this historic mission for a time such as this that most people ignore!

Can we use mathematics to scrutinize the Koran and prove that Islam is scientifically incorrect? Yes, and this book and others I wrote prove that. Can anybody really dare to say that? Yes, and I did so that people can understand.

25. Why did all the creationists fail to unravel the formula of the universe creation for so long?

"We cannot hold a torch to light another's path without brightening our own." — *Ben Sweetland*

"Do the hard jobs first. The easy jobs will take care of themselves." — *Dale Carnegie*

"Your attitude towards failure determines your altitude after failure." — *John Maxwell*

The simple answer is that, because most believers say they believe in God, it does not mean that they all got the Genesis story right in their hearts and brains, so they could mix it with their logic to unearth the rational explanation of creation that smart freethinkers, including atheists, will like—or won't they? In fact, many believers embraced evolution, which is equal to saying that God created nothing. As absurd as it may sound, some believers even went further to say that God used evolution to create the universe and its content. Before my discovery, most creationists thought and still think that science cannot explain the Biblical account of creation. Even after I demonstrated creation scientifically, countless creationists and their leaders could still doubt the literalism of the Genesis account of creation. In fact, when I broke the news of my discovery, countless top creationist leaders denied it.

Therefore, as they have already done to me, some creationists have been attacking God and all scientific theories, ignoring that even some wrong scientific theories contain raw data that can help formulate great hypotheses using some data contained in the Biblical creation, which the collectors of the raw data deny. Even today, some creationists continue to accept the billions of years theories, which blatantly oppose the literal explication of the days of creation. Some justify that by saying that Moses, who wrote the Biblical creation story, was not smart enough, or that the God of Israel, the Almighty and Omniscient, was not clever or scientific enough! How can people (even theologians and famous preachers) deny or believe that God created the universe in 6 days, as the Bible says, yet also believe that the days of creation were not 24 hours each? Worse, they act as if the atheists and freethinkers, who are unfortunately seeking physical proof of spiritual things, are the only enemies of God?

My contribution to all this scientific and theological mess is that my God-given life has prepared me to view things differently. I was able to even look for and separate great raw data from wrong scientific theories, and filter them through a unique perspective that most people would consider foolish. After 12 years of sacrifice and struggle (I cannot explain here), I came to a shocking conclusion: There is a God who created the universe in a matter of days!

But are Christians really demonstrating creation as they say? The simple answer is NO. Are Christians denying the God they want atheists to accept? The answer to that question will shock you! Before I explain that, let me remind you of this. It is easy for human beings to "*look at the speck of sawdust in their brother's eye and pay no attention to the plank in their own eye.*" When it comes to cosmological discussions through the lens of science or faith, it can be even easier for people to blame others for things they cannot see or prove. For instance, it is no secret that most Christians and non-Christians don't agree on many things, including the origin of the universe, yet they delight in blaming one another. Although most Christians can quickly say that non-Christians don't believe in God, can we judge without any doubt that all Christians really believe in God? What if countless Christians are denying or questioning God on critical topics just as some non-Christians do? Using the thorny topic of the origin of the universe, I will scientifically show you whether any group of Christians is making similar mistakes that most non-Christians do. So, let me ask you again: did Christians prove creation as they say? Before my discovery, they did not, and afterward, some Christians continued in the wrong direction.

What went wrong with the existing Biblical creationist theories? Hoping that the creationists will correct their theories so unbelieving freethinkers can discover and enjoy the God they hate, I cannot finish talking about creationists without specifically addressing each creationist group. I will first talk about the Old Earth creationists, then the Young Earth creationists, and the Intelligent Design proponents. I provided more information about these problems in *Science180 Accurate Scientific Proof of God* and *Reconciling Science and Creation Accurately.* I also devoted two books exclusively to children ages 7-12 and their parents:

- *How Baby Universe was born* (this is for unbelievers)
- *How God Created Baby Universe* (this is for believers)

Beginning with the Old Earth creationism, let me now address each creationist variant.

26. A sure way for Old Earth creationists to make mistakes and miss the historic moment— And how to avoid it?

"It is impossible to account for the creation of the universe without the agency of a Supreme Being." — George Washington

"Learn to say 'no' to the good so you can say 'yes' to the best." — John Maxwell

"No one but myself can be blamed for my fault. I have been my own greatest enemy—the cause of my own disastrous fate." — Napoleon at St Helena

Of all the Biblical creationist theories, Old Earth creationism (e.g., the creationist doctrine that claims that the days of creation are billions of years) can be put in the hot seat quickly. The scientific theory I demonstrated in this book shows that the Earth, the Moon, and the Sun were really formed in literal days, as the Bible says. But can we be right if we start accusing like Old Earth creationists? But why do Old Earth creationists think that God used evolution, the Big Bang, or other billions of years process to create everything, yet in 6 literal days?

If I cannot blatantly tell the Old Earth creationists that they make a big mistake that they should quickly fix, rather than continuing to justify what could be called nonsense, then, how can I even convince the atheists and non-Abrahamic believers that they are wrong? Without any delay, if they really love God, all Old Earth creationists must abandon their creation theory immediately, not because they are not Young Earth creationists. As you will see very soon, even Young Earth creationists have their share of the problem. And everybody must adjust if we really love the truth.

In summary, although Old Earth creationists confess that they believe in God (and I cannot judge them fully, for only God can say that), I can say that they completely missed the Biblical creation framework, because they deny the inerrancy of the timeline of the formation of the celestial bodies. And without embracing the literalism of the Genesis story, they will never refine their framework, which they need to throw away if they don't want to pay a high price.

Nevertheless, despite their ignorance of the depth of the Biblical account of creation, many Old Earth creationists believe in God and are saved. Therefore, nobody (even not the Young Earth creationists) should consider the Old Earth creationists as the worst, for God is the final Judge on how human search for knowledge (that God told Adam and Even to avoid) and their efforts to apply it to interpret the Scriptures can cost them heaven of the position they will occupy if they can get there by faith.

However, now that the truth is known, Old Earth creationists will have no more excuse to continue with their wrong creationist theories. They need to abandon all versions of the Old Earth creationism IMMEDIATELY, seek forgiveness from God right away, and embrace the truth regardless of how much it can cost them. Otherwise, they are in big trouble that they may not understand yet. No more excuses!

It is possible to liberate yourself and others from trying to present or understand the Biblical account of creation within an evolutionary framework. I can help you to enter a new area of freedom and power that crushes the head of all theories that have held the Biblical account of creation captive for so long. I can help you better resonate with your target market or customers who are craving something original that challenges anti-Biblical explanations of the origin of the universe and life, so you can increase your ability to create messages about the universe's origin and God's existence that attract and convert even the most stubborn unbelievers, increase your audience's engagement, cut through unbelief and resistance, and receive authoritative advice on the origin, functioning, and fate of the universe. It is possible to satisfy your audience's burning desire for freedom from beliefs and scientific theories about the origin of the universe and life that choke them and bind their minds, faith, hearts, and education. Do you want to know how to crowd cosmopolitical mistakes out of your mind, religion, and science so you don't regret tomorrow? If you do this, you will never miss the origin of the universe and the scientific proof of God. Contact me at www.Israel120.com to get the help you deserve.

Now, let's address the Young Earth creationists.

27. What to do when Young Earth creationists and Intelligent Design proponents cannot really claim victory or vengeance?

"The first responsibility of a leader is to define reality." — *Max De Pree*

"The Bible is true and science is true, and therefore each, if truly read, but proves the truth of the other." — *Matthew Fontaine Maury*

"Science affects the average man and woman in two ways already. He or she benefits by its application driving a motor-car or omnibus instead of a horse-drawn vehicle, and being treated for disease by a doctor or surgeon rather than a witch." — *John Burdon Sanderson Haldane*

"You can make more friends in two months by becoming interested in other people than you can in two years by trying to get people interested in you." — *Dale Carnegie*

Although Young Earth creationists and Intelligent Design proponents may be tempted to claim a victory lap, I would caution them not to do so too quickly without a proper retrospective on their methodology, some of their mistrust of science, and their embrace of what freethinkers (and even I may at one point) call pseudoscience. Indeed, Young Earth creationists are right in holding onto the literal interpretation of the Biblical creation, but are they really right in their approach?

Because most Young Earth creationist leaders don't think science can demonstrate creation, they focus on things not directly mentioned in the creation story. Hence, their frequent references to stories about dinosaurs, the Grand Canyon, Noah's great ark, fossils, etc., while they neglect the raw scientific data hidden in the Biblical account of creation. Furthermore, many Young Earth creationists see secular science as their enemy or enemies of God, just as the Democrats and the Republicans see each other as enemies and themselves as the holders of the ultimate truth, yet both are making similar mistakes they justify using biased/incomplete political frameworks that are not always putting God first (as they say), or making God known throughout the whole world (as God wishes), without raising unjustly, selfishly any political party, organization, country, or even race above the others. Nonetheless, the real enemy is spiritual and has no political party! Don't all of us benefit from many things that modern science has provided, including education, technology, and jobs, that some creationists betray themselves by embracing Big Bang and evolutionary concepts that they know (I assume very well) oppose the literal interpretation of the Bible. Can science, even secular science, be so evil and still profit us? The mindset that causes some people to perceive the relationship between science and religion as oil and water is also part of the reason some Young Earth creationists (and many other believers) have not achieved what I did. We need to be careful in our generalization of our experience. We should extract only the specific, relevant lesson from our experience, even with secular science, rather than applying it as a broad, irrational rule to think that science and religion are at war!

In short, among many reasons, Young Earth creationists were unable to demonstrate the inerrancy of the Genesis story scientifically in the eyes of the freethinkers because, just like the unbelievers, most of them think that such a demonstration is impossible to a point that their demonstrations focus not on HOW creation happened in 6 literal days, but on WHY they think it happened or "Why Design in Nature" points to God, that some Intelligent Design proponents strategically termed "Designer", which they say is not God, hoping that, by doing so, their demonstrations would be accepted. In other words, most proponents of Intelligent Design don't even like to be labelled as creationists, although (to my understanding) they intend to prove God as the Designer. But did the secular journals really embrace Intelligent Design? No. Yet, their intention is correct, but the method was wrong.

What is the point of trying to explore design in nature if, despite the critics, we cannot publicly testify that we attempt to prove God's existence through natural means? In short, Intelligent Design got the Biblical framework right (if they really embrace the literalism of the Genesis story). Still, their scientific framework is not as bold as that of Young Earth creationists (which lacks the methodology). As they embrace Science180 Creationism (see www.Science180Creationism.com) I spearheaded, they will find easily the design and the Biblical inerrancy answers they have been looking for scientifically in the creation story.

In other words, Young Earth creationists believe that God created the universe in 6 literal days. Still, many of their leaders perceive secular science (which contains the raw data that should be used to demonstrate what I did) as evil and the enemy, which some refuse to fully acknowledge as spiritual and located beyond political parties or national borderlines! With all humility, I believe and I can prove to you that some staunchest defenders of the Biblical inerrancy have been unable to free great raw data from wrong theories and use them to forge new scientifically verifiable theories that can better help atheists and all other freethinkers to rationally understand and embrace the God they hate, which most believers want them to accept while leaving their rational mind at the door! We need to think beyond the era of so-called blind faith. For we now have the proof for faith, and we need to embrace it so we can win souls for God, rather than judging them and chasing them away with the newly discovered truth or proof. Once freethinkers get the scientific proof, many of those who deny God vehemently can finally know the importance of faith. Hence, I know that my discovery has opened a new door that will advance humankind and bring peace of mind at the crossroads of science and faith!

Hence, I am grateful to God for having helped me do boldly what people thought was impossible. Therefore, despite the opposition I am facing from both unbelievers and even believers, because I challenged the status quo, I will never stop talking about the scientific and mathematical proof of God's existence at the intersection of science and faith! To learn more about my candid analysis of creationism, please consult *Science180 Accurate Scientific Proof of God, Reconciling Science and Creation Accurately*, and others in the pipeline.

In summary, although Young Earth creationism and Intelligent Design got part of the interpretation of the creation story right, that does not mean that their proponents are the best or right on every issue pertaining to creation. They still have homework to do to correct their story so freethinkers can enjoy God. This also explains why Science180 Creationism, or, let's say, Science180 Cosmology, is so unique: it allows people to discover God, even while keeping their reason. For faith is not irrational, and reason is not against God, who created the mind and gave it to human beings! But be sure to believe the right way and use your reason right.

I can help you understand with certainty how to present the scientific evidence in the Biblical account of creation that you have been seeking, and liberate yourself and others from trying to present or understand it within an evolutionary framework that is false. With my discovery, you can scientifically explode the myth that the days of creation in the Bible were millions of years long rather than 24-hour days. I can help you discover how modern science proves the inerrancy of the Biblical account of the creation of the universe so that you can put a full stop to all (scientific and religious) efforts that threaten to undermine faith. By doing so, you will avoid regrets in the end, stop wasting time on useless efforts attempting to convince the staunchest rationalists, start focusing on the real answers that freethinkers need before embracing the God they hate, reach, and convert more prospects into profitable and loyal believers, boost your margins of profits, and outpace your competitors. Do you want to know how to break, destroy, and tear apart all evolutionary theories about creation before they destroy you? Check out more about the scientific breakthrough that will outlaw many of your cosmological challenges at www.Science180Academy.com.

28. What went wrong with atheists, agnostics, and all other freethinkers?

"Many people will tell you that an expert is someone who knows a great deal about the subject. To this I would object that one can never know much about any subject. I would

much prefer the following definition: an expert is someone who knows some of the worst mistakes that can be made in the subject, and how to avoid them." — Werner Heisenberg

"Any time scientists "try to fool Mother Nature by averaging out her details, she gets her revenge by leaving them with a closure problem — more equations than unknowns." — William K. George

"Do the thing you fear to do and keep on doing it—that is the quickest and surest way ever yet discovered to conquer fear." — Dale Carnegie

"We may ignore, but we can nowhere evade the presence of God. The world is crowded with Him. He walks everywhere incognito." — C. S. Lewis

Why were atheists and all other freethinkers not able to discover my theory? Not only did they not discover it, but also, considering what I went through during the 12 years leading up to the publication of my discovery, I can report that they would never have discovered it using the modern methodology of scientific research that makes people (particularly unbelievers) think that all religions are unreasonable, or that all religions lead to God—neither science or religion can save you, only God (Jesus, Yeshua HaMashiach) can! Because most freethinkers (including atheists and agnostics) already deny or doubt God, they would have never found by themselves some key secrets to unlock key variables and refine the formula I discovered. Although most freethinkers claim they are looking for rational proofs, I have come to discover that most people are biased even when it comes to the proof they claim to want. When I was trying to break the news online through press releases, I was told repeatedly that the media does not want anything to do with God, and worse, anything attempting to prove God! Hence, the freethinkers need to be willing to admit their share of the global mistake and, after carefully reviewing the facts I presented here, to change their minds when they find a mistake in their logic.

If I could detail how I refined this formula here, you could know that those who start with the premise that there is no God or that science can never prove God, or that religious people are irrational, or that science is against God will never discover the formula that actually not only proved God's existence, but that also proved how He operates and wants us to find Him through spiritual means including faith, which unbelieving freethinkers think is irrational as if everything they do in science is rational. I wish I had time to elaborate on the irrationalism of modern science and the belief credo in the modern methodology of scientific research! But that's a topic for another time!

I understand why most unbelieving freethinkers (e.g., atheists, humanists, rationalists, agnostics, nontheists, nonreligious, skeptics, nonbelievers, religiously unaffiliated, spiritual-not-religious, ex-believers, and doubters) think that there is "no rational justification" for believing in God. Hence, some of them consider religious people to be mad, psychotic, and delusional. But, considering at least the evidence I presented here, is that really true?

Isn't it also said more than 100 years ago that extreme skepticism and blind credulity are intellectually "lazy" because they allow people to avoid the hard work of critical thinking?:

"To doubt everything or to believe everything are two equally convenient solutions; both dispense with the necessity of reflection." — *Henri Poincaré*

I hope freethinkers can logically examine the existence of God, not by just agreeing that excellent things in nature prove God's existence, but by checking the tangible, scientific facts I offered here, which they can rationalize in their mind, so they discover how science does not threaten or undermine faith, but confirms the Biblical account of creation as true, even when we now know that previous creationists have failed to prove creation scientifically in the eyes of the unbelieving rationalists.

I can help you to examine the crossroad of science and religion critically so that you open new, groundbreaking doors to ask new questions focusing on the real origin of the universe and the existence of God, cut useless costs, stop wasting time on useless ideas, understand the past of the universe, innovate today, and prepare for a better future. He who can understand this has the scientific code of the universe's formation with him. If you master this, you will never need anyone else to explain to you the origin of the universe or the scientific proof of God's existence. He who cannot or doesn't want to will stay behind. If you are a freethinker, what is still making you doubt God with the scientific evidence? If you want me to help you address it, please get in touch. I will help you free of charge. Check out www.Science180.com.

But what really went wrong with evolution and the Big Bang theory? Let's talk about that quickly.

"It is not the strongest of the species that survive, nor the most intelligent, but the one most responsive to change." — *Charles Darwin*

29. He who can understand why evolution and the Big Bang theory missed the universe's origin formula has the key to his destiny. He who cannot journeys toward his own perdition

"What I am going to tell you about is what we teach our physics students in the third or fourth year of graduate school... It is my task to convince you not to turn away because you don't understand it. You see my physics students don't understand it... That is because I don't understand it. Nobody does." — Richard P. Feynman

"The greatest discovery of all time is that a person can change his future by merely changing his attitude." — Oprah Winfrey

"People who boast about their I.Q. are losers." — Stephen Hawking

"We always overestimate the change that will occur in the next two years and underestimate the change that will occur in the next ten. Don't let yourself be lulled into inaction." — Bill Gates

"There is nothing permanent except change." — Buddha

"The only constant in life is change." — Heraclitus

When I first submitted my theory of the origin of the universe to scientific journals in 2025, I labeled it Science180 Turbulence 1, in remembrance of the mother of all turbulences I proved shaped the universe. As I talk about scientists in this segment, I will refer to my theory again as Science180 Turbulence 1.

Undoubtedly, all secular theories about the origin of the universe and life (e.g., evolution and the Big Bang theory) have rejected God from the equations. Until the publication of my book, the top scientific journals didn't publish much (if anything) about the Biblical account of creation. Although some evolutionists would defend that they believe in God, even the God of the Bible, who says that He created the universe in 6 days, it is not clear whether they were mistaken. Embracing billions-of-years theories wholeheartedly can be plainly denying God the creator.

Without a mindset that knows modern cosmological theories were wrong, without knowledge of the Biblical account of creation, and without a desire to follow the raw scientific data and allow them to lead me wherever the truth is, I could not have ever discovered the truth behind creation scientifically. Even so, some people will continue to debate what the TRUTH is and what is Fake! Well, forgive me for choosing not to argue that now.

Therefore, secular scientists failed to find the compass that could have led them to the intersection of science and faith, where the real understanding of the origin of the universe began. I am not shocked that some of the top scientific journals to which I submitted my original God-related articles refused to send them to reviewers. You can learn more about that in my book '*Mathematical Proof of God's Existence at the Intersection of Science and Faith.*' They knew my discovery was groundbreaking, but they probably fear for their careers, their interests, their egos, and their framework they will have to rebuild, whether they like it or not. For the truth is finally revealed scientifically. And now is the time to know which religion is rational and which science is irrational! And when people discover the truth plainly, nobody, not even a department of education or school board of education, can stop them from trading ignorance for true knowledge and even for the faith that can be finally demonstrated scientifically. That is why I know that all other religions and even scientific endeavors that oppose God are in trouble, for well-informed people will no longer bow to irrational theories, myths, and fiction that have been fighting to raise themselves above the knowledge of the highest truth, which we can never find without God the creator, who is not limited by time, space, or political parties or national borderlines!

Although science has helped humankind tremendously, I must admit here that evolution, the Big Bang theory, and other secular theories have missed the big picture in their efforts to decode the origin of the universe. Instead, they used wrong frameworks and narrow datasets collected on the universe and life to craft weak theories that, unfortunately, have been dictating what the scientific community must follow, ignoring that their strategy has tied the truth and opened the gate for lies (that science cannot test or disprove only by itself), which we now know must be abandoned—or don't you think so? Otherwise, science could become a religion (if it is not already, in the view of many believers whom most secularists perceive as irrational), and Judeo-Christianity would become the supra-science to which people could turn to make peace with their faith and their search for rational knowledge. Or is that precision irrational?

Most people don't want to hear this, but, considering the implications of my discovery and other predictions, that's the hard truth I cannot afford to sugarcoat. Only the Biblical account of creation, told thousands of years ago, confirmed the scientific theory of the formation of the universe that I built using data collected by top scientists throughout the ages!

To wrap it up, if you ever asked yourself whether the Bible can save science (including theories like Big Bang and evolution), even when most people think science and the Bible are at war, then the answer is a big YES. But without sticking to the Biblical account of creation, science cannot find the truth needed to reorient itself in our era, where artificial intelligence is helping create and misanalyse massive amounts of data, before using its conclusions (true or false) against humankind more than the leaders of that technology would confess. This may not be completely true, but that is how some believers perceive this AI fever, let's say pandemic! But are all believers attacking secular scientists ungrateful to science? Let's check it out next.

"Change is the law of life. And those who look only to the past or present are certain to miss the future." — John F. Kennedy

"The price of doing the same old thing is far higher than the price of change." — Bill Clinton

"Faith is a walk — a journey toward greater understanding … You are equally an American if you choose to worship an almighty and if you choose not to." — George W. Bush

"Change will not come if we wait for some other person or some other time. We are the ones we've been waiting for. We are the change that we seek." — Barack Obama

"We choose hope over fear. Unity over division. Science over fiction. Truth over lies." — Joe Biden

"America needs God now more than ever." — Donald Trump

"In God We Trust." — Mandatorily printed on all American coins and paper money since 1966

"Ignorance more frequently begets confidence than does knowledge. It is those who know little, and not those who know much, who so positively assert that this or that problem will never be solved by science." — Charles Darwin

30. A sure way to be ungrateful to modern science! How to avoid it

"Failing forward is the ability to get back up after you've been knocked down, learn from your mistake, and move forward in a better direction." — John Maxwell

30.1. Is it difficult for people to blame modern science, not themselves?

"Two things are infinite: the universe and human stupidity; and I'm not sure about the universe." — Albert Einstein

"Any fool can criticize, condemn and complain—and most fools do." — Dale Carnegie

Considering my discovery, it will be easy for some people to blame modern science and to say that science as a whole missed the theory. But is that true? Although modern science has missed the scientific formula of God and the creation of the universe, not everything in science is evil, just as not everything in all religions is good.

For centuries, science has operated within a framework that has excluded God. Yet science has helped tremendously in addressing some global problems. But the new theory that I introduced in this book and others I wrote (see www.Israel120.com/books) shows that we (scientists and nonscientists, unbelievers and even believers) were missing a huge part of the big picture. Because science is a process of constant refinement, now is the time for the real scientific work to begin: to understand Science180 cosmology so we can save lives more effectively. Hence, as I thank scientists for their efforts throughout the ages, I also need to warn them of the danger of some current scientific models, and we will pay a higher price if the leaders cannot admit their mistakes and correct them quickly. Those who will respect my discovery will take the lead as history will prove.

Why did modern science fail or miss the theory then? Because scientists rejected the truth too early, which could have recalibrated them, and those who believe in the truth did not have the perception or the guts to abandon some useless secular theories, so they could be enlightened to discover the needed alternative theory! Furthermore, most scientists embraced theories that seemed true, while rejecting theories that were not physically accessible at the time. But can we prove that most people will even embrace the Bible? By trusting the work on men over the creative work on God, we ended up trading priceless truth and the invisible reality against misleading physical correlations and fiction we taught ourselves to believe against the reality we cannot prove scientifically YET. And we call that intelligence! And when we transfer that mindset to our gadgets, we call it artificial intelligence. Though we know it is artificial, we still use it because it is natural to crush what we think won't profit us!

By the way, after my discovery, I encountered some people who claimed they were seeking rational proof of God before believing in Him, but when the scientific proofs were presented to them, they said they don't publish, read, like, or share anything like that. I met those people not just in the secular world, but even in the church, among the most zealous people you may think of. Whether they are educated or not, some of those people are leaders in academia, media, AI industries, preaching platforms (clergy), and many other influential places, where the truth is twisted and then presented to a naive crowd that loves fiction or imaginary things more than non-fiction dealing with reality. When I put all these facts together with other observations, I knew that AI would be one of the main enemies of humankind. Still, humankind will love it, ignoring that naively consuming AI is like drinking of the most dangerous poison!

Anyway, as you will read in my upcoming autobiography, you will know how what I went through in life has prepared me to do what nobody on this planet knew was possible. Then, you will understand why no scientist before me was able to crack the code of the formation of the universe.

To illustrate what I just said about the failure of modern science, I will explain why scientists missed the implication of 4 key variables I used to crack the universe's origin. Indeed, I am not the person who invented or collected the radius, orbital speed, escape velocity, and the distance related to the Earth, the Moon, and the Sun. But I am the one who proved for the first time in history what really happened during the formation of the universe, so those 4 variables could be set the way they are! I will now elaborate on two things:

- Why my theory would have left Albert Einstein speechless, and Isaac Newton puzzled
- Why would the mathematician Leonhard Euler (I will talk to you more about who he is very soon) have been shocked that he missed my formula?

30.2. Could Albert Einstein have been speechless, Isaac Newton puzzled, and Leonhard Euler shocked by the mathematical demonstration of Science180 Turbulence 1 that they missed in plain sight??

"Equipped with his five senses, man explores the universe around him and calls the adventure Science." — Edwin Powell Hubble

"Knowledge of physical science will not console me for ignorance of morality in time of affliction, but knowledge of morality will always console me for ignorance of physical science." — Blaise Pascal

Although physicists and engineers knew about the escape velocity (one of the main 4 variables I used in my theory) since the days of Isaac Newton, they did not understand its cosmological importance. One aspect of my contribution is to demonstrate how this famous variable can be incorporated into the methodological framework to reconstruct the precursors' journey and the formation times of celestial bodies.

But why did scientists before me miss it? Many reasons can be cited, and I have detailed them in my books. But here, I will mention just a few. Indeed, since most physicists think that the current position and movement of celestial bodies is random, it never crossed their mind that, by paying attention to the distance separating the bodies and their speed, they can actually know how and when they were formed. Therefore, they created wrong references and mechanisms to study the Solar System and the universe as a whole. For instance, scientists thought that the bodies orbiting the Sun began their journeys in the region of the giant planets (e.g., Jupiter and Saturn). In other words, before my discovery, scientists reasoned that the precursors of some celestial bodies in the Solar System moved from the region of Jupiter toward the Sun, whereas others moved outward from that region! Hence, they wrongly thought that a certain asteroid had hit Earth, producing the Moon as debris from the alleged collision. They did not know that the Earth and the Moon descended from a common ancestor, which I called the precursor of the Earth-Moon System, which, in turn, descended from the precursor of all the bodies orbiting the Sun, which, in turn, descended from the precursor of the entire Solar System. They also have no clue about the precursor to the bodies organized into fluid layers, and this misunderstanding led them to think that it took billions of years for the layers of the Earth's crust to form. In the view of these scientists, Earth's layers were deposited gradually, and by positing the rate of deposition, they thought they could determine the age of the universe and the processes that were involved. It was the misunderstanding of the existence of fluid layers in the precursors of celestial bodies, which were then wrapped around to form them, that caused scientists to miscalculate the age of the universe by billions of years. For instance, they thought that particles in the Earth's layers were deposited at a certain speed or rate, and by dividing the depth of the Earth's layers by that alleged speed or rate, they arrived at a number purported to determine the age of the Earth's formation. Wrong! I deal with that in another book.

If scientists had really understood the meaning of the escape velocity in the context of the precursors of the celestial bodies orbiting the Sun escaping the Sun, they could have put themselves on the right path a long time ago. Sir Isaac Newton, whose work led to the calculation of the escape velocity, did not imagine this scenario because, among many things, he thought that the Sun was attracting the other bodies in the Solar System. Hence, he could not imagine that, during the formation of the Solar System, the precursor of the bodies orbiting the Sun escaped the precursor of the Sun, which, in his mind, must attract everything orbiting it. Likewise, Isaac Newton did not imagine that, during the formation of the Earth-Moon System, the precursor of the Moon escaped the precursor of the Earth, because his theory held that the Earth and the Moon attract each other. When Albert Einstein came centuries after Isaac Newton, he mathematically proposed the notion of the curvature of space (which is false, by the way) as an alternative to the gravitational theory that Sir Isaac Newton postulated. In short, both Sir Isaac Newton and Albert Einstein did not understand the meaning and origin of gravity, which, as I explained in my other books, are closely connected to the split-gathering process I discovered. And because modern cosmology is mostly based on the classical physics of Sir Isaac Newton and the relativistic foundation of Albert Einstein, scientists have been in an intellectual bondage that has prevented them from unlearning wrong things they were taught or taught others. In my book *Turbulent Origin of the Universe* I detailed that issue.

Isn't it shocking to you that physicists knew the escape velocity for almost 400 years, but it never crossed their minds that the precursor of the Earth could have escaped the precursor of the Sun before arriving at the orbit of the Earth! Discovering my cosmological perspective on escape velocity was not an easy task for me. It took me years of complex analysis to decode this piece of my theory. But how the scientific community is already fighting me because of my discovery tells a lot about how facts and logic have lost their meaning (if a real meaning even really ever existed in their mind) at the expense of some scientific prejudices we fail to label as religion!

Anyway, Sir Isaac Newton could also have discovered my formula if, among many things, he was not so attached to his ideas about gravity and how celestial bodies attract one another. Albert Einstein could have discovered the formula if he had not let mathematics overshadow his understanding of the so-called attraction, which Sir Isaac Newton had gotten wrong. Because math was making him think that space is curved, Einstein (a Jew who should have known the Genesis story since he was a kid) threw away the Biblical account of creation that some of his ancestors treasured so much that he did not pay much attention to the God of Abraham, Isaac, and Jacob. Therefore, he neglected the Biblical truth that could have guided him to refine or abandon part of his relativity theory, which countless scientists today treat as gospel, ignoring its mistakes! Having found a mathematically acceptable alternative to Newton's law of gravitation, Albert Einstein thought he had found the ultimate answer, rationally devised as a relative theory. For more than 100 years, the scientific community believed that Einstein's theory of relativity was perfect, yet they knew it had mistakes! Therefore, scientists brag all the time, saying that everything is relative, while ignoring the existence of an absolute truth, just as Albert Einstein did.

> *"I cannot imagine a God who rewards and punishes the objects of his creation, whose purposes are modeled after our own -- a God, in short, who is but a reflection of human frailty. Neither can I believe that the individual survives the death of his body, although feeble souls harbor such thoughts through fear or ridiculous egotisms."* — *Albert Einstein*

BOLDEST SCIENTIFIC FORMULA OF GOD AND CREATION

"The word God is for me nothing more than the expression and product of human weaknesses, the Bible a collection of honourable, but still primitive legends which are nevertheless pretty childish. No interpretation no matter how subtle can change this."— Albert Einstein

"It seems to me that the idea of a personal God is an anthropological concept which I cannot take seriously. I also cannot imagine some will or goal outside the human sphere... Science has been charged with undermining morality, but the charge is unjust. A man's ethical behavior should be based effectually on sympathy, education, and social ties and needs; no religious basis is necessary. Man would indeed be in a poor way if he had to be restrained by fear of punishment and hope of reward after death." — Albert Einstein

If I can list some anti-creationist quotes credited to Albert Einstein here, some Jews who lived 2000 years ago, and who accused Jesus Christ as someone blaspheming God, might have stoned or try to stone him to death, if they could apply the same rigor or zeal they used to mistakenly interpret the Hebrew Bible, which some may say is as strong as the enthusiasm some secular scientists use to reject creationist facts at the profit of a billions-of-years process they think really formed the universe.

This may sound funny or weird, but those facts I must deal with thoroughly (as I debunk failed theories) in my candid analysis intended not to please or fear anybody, but to set people free from all forms of intellectual and religious captivity related to the formation of the universe, so from now forth, all realistic cosmological debates can no longer exclude the Creator, whom we accused for almost 6000 years, beginning with Adam and Eve, whom thought that, by disobeying God and eating the forbidden fruit of knowledge, they could become like God (just like Satan deceived them to believe). Today, some people are embracing AI and other technologies (whether biotechnology or Silicon Valley technology) to try to engineer human beings, whom (they allegedly say) will know a lot and act like robots (by the way, they won't tell you that) as if human beings were not created perfect, or as if the solution of our problems (most of which we create ourselves, while saying that the Creator created nothing) is still the anti-God knowledge, even the knowledge (not just some secular knowledge that some believers hate, but even some religious ones as well) that denies the truth we hate at the expense of the (Big) lies and fiction we embrace naively—all this because we don't know our real problems, needs, and fate! Please don't misunderstand me: I am not against technology, and I am not saying that everything about technology is bad, but when we deny the merciful and almighty God and naively embrace technology that causes us to think that we can engineer or become better human beings or better living things than the ones created by God, that we deny, I don't need anybody else to tell me where this whole technological revolution is going. As I know that my discovery will open the eyes of scientists to interpret better the massive amount of data collected before me, I must warn the entire scientific and engineering community about how certain things they are running after (for the sake of money, fame, or power) can go bad, even from bad to worse, overnight. Therefore, I need to wash my hands from any nonsense they will do with my discovery, for which I don't want to be part of!

Why would the mathematician Leonhard Euler have been shocked that he missed my formula? Leonhard Euler (1707-1783) can be considered the father of the scientific formula that allows proper calculation of the relationship between the speed of celestial bodies and the distance separating them from the bodies they orbit. He could have found my theory if he had just imagined for a moment why the speed of the planets decreases as they are farther from the Sun, or how the precursor of the bodies orbiting the Sun escaped the precursor of the Sun. But things are not as easy as you may think. I told you earlier that, until my discovery, scientists thought that the bodies orbiting the Sun moved from the region around the giant planets (such as Jupiter and Saturn). So, they never imagined that, in fact, the precursors of the bodies orbiting the Sun moved from a position near the Sun to the outskirts of the same.

Therefore, although there are trends in the orbital speeds of bodies orbiting the Sun, their perspectives never allowed them to unravel key components of the formation of the universe. Do I have to inform or remind you that the distance between the Sun and the Earth, and that between the Moon and the Earth, were known to an acceptable accuracy long before Jesus Christ was born about 2000 years ago! But why has nobody ever thought about using those numbers to formulate some hypothesis about the real formation of the universe?

"Science without conscience is but the ruin of the soul." — François Rabelais

"Common sense, 'the collection of prejudices acquired by age eighteen."' — Albert Einstein.

I did not sit down and imagine scenarios to put these 4 variables to work. I used a unique inductive method of analyzing countless data, meaning that as I investigated countless celestial bodies, patterns and relationships between previously unrelated variables emerged. It took sacrifice and hard work (without any human being or institution renumerating me) to get there, something that most scientists and even nonscientists won't accept. You will learn more about that very soon.

Well, more than 2,000 years ago, the Greek scholar Eratosthenes (an African) calculated the Earth's radius by analyzing shadows. Now, another African, let's say Beninese-American (Nathanael-Israel Israel), is telling you how the entire universe was formed. Will you show some respect to this unconventional Black man and listen without letting any prejudice bias you? As Nathanael told Jesus Christ, "*Can anything good come out of Nazareth*" [the hometown of Jesus Christ], are you already asking yourself this question: "*Could any significant scientific milestone come out of a Black man (born in Benin, a rather tiny, mostly overlooked French-speaking country in Africa, which 99% of the Americans I met these last 20+ years, ignore), who SOME people think cannot be a thought leader?*" Can I respectfully ask them to remember that even Christ's disciples were initially skeptical that anything good could come out of Nazareth, His rather modest hometown, yet it did? Well, I know you won't ask that question, but if you do, you won't deny how those who will embrace this discovery will surpass you if they are careful! But why the italics in the previous quote? If you want to know, please contact me!

To wrap it up, although a breakthrough was needed in turbulence before the universe could be fully understood (and I made that breakthrough as detailed in my book *Turbulent Origin of the Universe*), scientists have turned their back against the truth that could lead them to ask the proper questions and know how to combine the existing data (not just in science) but also in religion (which some scientists think is illogical or ridiculous), so they could have the key to unlock the origin of the universe that no one can never access without a certain mindset, which conventional scientific method can never provide! Hence, it took an outside-of-the-box scientist and thinker like me, coming out of nowhere, overcoming countless challenges, to shock the whole world! Most people don't want to let light come out of the darkness they sometimes find themselves in. I wish I could tell you about the uniqueness of my life story right now and how I was prepared to perceive things differently from most people, even at the expense of my "comfort," if I even ever have any or looking for any such a thing in this complex world (where I am supposed to live like a sheep among wolves) that is about to end—or am I dreaming? Should anyone even make it in this life by relaxing in his or her Kumbaya Eldorado Comfort Zone (please don't ask me what that means or where you can find that paradise on this planet Earth)? But don't worry, you will figure it out very soon. But how can others use my discovery? Do you want to take advantage of the magic formula that can help you solve the biggest cosmological problems? Do you want to know how to break, destroy, and tear apart all nonsensical theories about the universe's origin before they destroy you and science? Do you want to innovate in your field?

I can help the members of your organization become the leaders, scientists, engineers, or innovators who capture the hearts of their followers, prospects, and customers, craving an unconventional yet accurate explanation of the origins of the universe, life, and chemicals. I can assist you in positioning yourself to acquire the ability, insight, intelligence, wisdom, and extraordinary knowledge to interpret complex scientific data, explain riddles, and solve the most complex scientific problems related to the formation of the universe, and to apply them to your field. Taking advantage of the holy grail I discovered at the demystified intersection of science and faith, I can help you to craft new research projects, ask new questions focusing on real solutions, stop wasting time on useless scientific problems that will yield nothing, stop spending your resources on unprofitable projects but on profitable ones, take your organization to a higher level, beating your competitors. I can't wait to hear from you at www.Science180Academy.com.

"Do not be afraid of being free thinkers. If you think strongly enough you will be forced by science to the belief in God, which is the foundation of all Religion." — Lord Kelvin (William Thomson)

"Creator has arranged the world in such a way that nothing is left to chance" — Gregor Mendel

"Knowledge informs the mind. Wisdom transforms the situation. Knowledge fills you with facts. Wisdom gives you the solution. You may have degree, diploma, but still make foolish choices. Wisdom cannot be earned by study. It must be received by revelation. Knowledge enters through the eyes. Wisdom flows through the spirit. School may give you education, but only God gives you insight. The world admires intelligence. But heaven celebrates understanding. You can research for years. Yet remain blind to what is true. For wisdom comes not from effort, but from encounter. It is when the spirit touches your heart that light floods your darkness." — David Yonggi Cho

TAKE-HOME MESSAGE OF SECTION 5: EVALUATING MY THEORY AND TROUBLESHOOTING FAILED THEORIES

As I evaluated my theory in light of the existing scientific and religious accounts, the scientific evidence bizarrely aligned with the Biblical account of creation. In other words, considering all the religious stories in the world, it is remarkable that, among all the religious books, only the Bible contains a creation story that matches the scientific evidence. As I troubleshooted the existing (scientific or religious) theories, I discovered that they failed because they are based on methodological frameworks. By paying attention to the code behind the Nathanael-Israel Israel Birthdate of celestial bodies, developed after my Turbulent Split-Gathering Framework, it becomes easy to test scientifically whether any religion in the world is incorrect. With that method, I use mathematics to scrutinize the Koran and prove that Islam is scientifically incorrect. The same is true for all other religions except Christianity and Messianic Judaism. Still, the creationists (even the unshakable ones) failed to unravel the formula for the creation of the universe for so long. Whether they are Old Earth creationists, Young Earth creationists, or Intelligent Design proponents, errors abound to the point that none of them can really claim victory with their existing methods. I showed what went wrong with atheists, agnostics, all other freethinkers, evolution, and the Big Bang theory. Nevertheless, I cautioned the world about the sure way to be ungrateful to modern science and how to avoid it, so we can reap the fruits of discoveries that will shake the world to improve life.

Section 6: Science180 Differentiators — Why is the Turbulent Split-Gathering Framework (the Methodological Framework of Science180 Cosmology) Novel, and Why Does it Add Value Compared to Existing Frameworks and Theories?

"The difference between appreciation and flattery? One is sincere and the other insincere. One comes from the heart; the other from the teeth." — *Dale Carnegie*

"Nothing is so embarrassing as watching someone do something that you said couldn't be done." — *Sam Ewing*

"It isn't what you have, or who you are, or where you are, or what you are doing that makes you happy or unhappy. It is what you think about it." — *Dale Carnegie*

"A new scientific truth does not triumph by convincing its opponents and making them see the light, but rather because its opponents eventually die, and a new generation grows up that is familiar with it." — *Max Planck*

"Whatsoever things are true, whatsoever things are honest, whatsoever things are just, whatsoever things are pure, whatsoever things are lovely, whatsoever things are of good report; if there be any virtue, and if there be any praise, think on these things." — *Paul the Apostle*

31. What is in this section for me?

"Physics isn't the most important thing. Love is." — Richard P. Feynman

"It isn't what you have, or who you are, or where you are, or what you are doing that makes you happy or unhappy. It is what you think about it." — Dale Carnegie

Most people spend more time planning their summer vacation than planning their lives" – – Source unknown

"Plato is my friend, Aristotle is my friend, but my greatest friend is truth." — Isaac Newton

Discover the truth and hold onto it: Remember, you have only one life, and you cannot play with your destiny if it is attached to your decision about creation and the scientific theory that will outlaw many cosmological nonsenses. To help you avoid confusion, I need to fence off my theory so it stands out from existing scientific and religious theories. Although many theories exist in the literature about the origin of the universe and life (the Big Bang theory and evolution being the most widely accepted in the secular education system before my discovery) and many creationist theories on the market, I want to emphasize that the theory I advanced in this book and others I wrote are unique because they are:

- An easy-to-understand universe-origin and God's existence theory
- The state-of-the-art decoding experience of the origin of the universe
- The most accurate, reliable, safest, and best explanation of the universe's origin ever
- The universe-origin theory that helps you fight wasteful programs and theories
- The universe-origin rule breaker
- All the universe-origin solutions you love
- Bringing the Judeo-Christian believers together through the power of the accurate decoding and understanding of the Biblical account of creation, so they can better help humankind discover their true origin and the meaning of life
- The undeniable reconciliation of science and the Biblical account of creation
- Capable of testing the falsification of the world's religions through the intersection of science and faith accurately

To learn more, visit www.Science180Creationism.com

32. Easy-to-understand universe-origin and God's existence theory

Most existing theories about the origin of the universe and the scientific proof of God are complicated, inaccurate, expensive, time-consuming, and labor-intensive. Whether you are a physicist, a chemist, a life scientist, a social scientist, or any other type of expert, and even layperson, you will find the theory I presented here and in my other books very easy to use in helping you take your career, initiative, research, curiosity, knowledge, and ability decode the universe's origin to the next level. Rather than dealing with incompatible and inconvenient theories, you can get a comprehensive understanding of the most crucial origin questions in one place. Grounded in empirical observations, this book targets the most challenging universe-origin problems and the scientific proof of God's existence from a fresh angle, using strategies that accurately break down the complex origin hurdles into a language that you can easily understand.

33. State-of-the-art decoding experience of the origin of the universe

This innovative theory creates an extraordinary experience for scientists, laypeople, children, and anyone else seeking an accurate understanding of the formation of the universe. Indeed, the proper universe-origin theory can make your life experiences enjoyable rather than give you headaches due to a false approach, outdated perspective, or unrealistic philosophy. With this unconventional theory, you can enjoy a comfortable, hassle-free experience that leads to an unprecedented understanding of how the universe formed. Without bothering you with unnecessary and irrelevant data, formulas, or theories, this theory is carefully crafted and presented with your origin-related needs in mind. It helps you clearly use a new scientific method that integrates various disciplines to decode how the universe's content came into existence. By helping you connect the complex dots regarding the formation of the universe, you will have a world-class experience of how the universe and its contents came to be. Instead of having to replace outdated universe-origin books with new ones very often (because they were incorrect), you will take pleasure in this innovative, well-thought-out, long-lasting, and time-tested theory, and the other products (books) and services (see Science180.com) connected to it.

34. The most accurate, reliable, safest, and best explanation of the universe's origin ever

Most theories about the origin of the universe and the existence of God don't address the critical questions about origin that arise across all fields of knowledge. Therefore, most existing theories are not comprehensive; they only partially fit a limited amount of data. That is why many universe-origin theories have been abandoned and replaced by new ones, which would be replaced shortly as well. In contrast to existing cosmological theories, Science180 Cosmology, which I broke down here, considers a broad range of data and uses a clear-cut approach that makes its universe-origin explanations reliable and applicable across many disciplines. It helps you to safely embrace its authoritative findings, knowing that they are scientifically proven to be correct. It has everything you would want to know about a premium, valid theory of the formation of the universe and the scientific proof of God. The complex data it examined and its unique perspective will help you to enjoy properly decoding, playing with, and mastering the universe-origin like never before. With an ultra-powerful ability to explain difficult concepts related to the formation of galaxies, stars, planets, satellites, asteroids, and other things in the universe, using simple terms and an unmatched accuracy, this theory will help you cut useless costs and time-consuming research related to your efforts to unambiguously understand the origin of the universe and life, while helping you to stop wasting resources on useless and unrealistic theories that will yield nothing. By doing so, you can focus your efforts on real problems that can make your initiatives more profitable and take you and your organization to a higher level, where new, groundbreaking doors will open for you. At the same time, you outpace your competitors who are still struggling to accurately understand the beginning of the universe and prepare for a better future.

35. Universe-origin theory that helps you fight wasteful programs and theories

This universe-origin theory is sustainably crafted, inexpensive, and designed to fit your needs. Planning all your projects can be overwhelming and resource-consuming, especially if you are trying to ensure no resources (including money) go to waste. Science180 Cosmology focuses on critical origin-related variables that most scientists have mistakenly discarded due to misunderstandings about the universe's true formation. Therefore, unlike all other cosmological theories, Science180 Cosmology debunked all existing scientific and religious theories on the formation of the universe, and will help you avoid wasting your resources on unprofitable origin projects so that you conserve more cash and spend it on more valuable and profitable things.

36. Universe-origin rule breaker

Science180 Cosmology or Science180 Turbulence 1 doesn't play by the ancient or modern rules set by some people who have tried to explain the origin of the universe using theories that refuse to acknowledge the real meaning of the scientific data. Instead, it used the data collected throughout the ages to scientifically decode the origin of the universe from a perspective that will surely help you, free you, empower you, challenge you, motivate you, and cause you to dance to a fresh and original beat that takes you many steps closer to your best life today and forever. Am I scared of what the established orders will say? No. Do I fear being controversial and original? No. Am I concerned about the useless theories that Science180 Cosmology will dethrone or destroy? No. Am I concerned about sticking out our tongue at traditional and secular universe-origin and life-origin nonsense? No. Am I concerned that some people may not like me? No. With Science180 Cosmology, I focus on my message and my mission, knowing that by doing so, I will help many people including you.

37. All the universe-origin solutions you love

People are tired of the overwhelming number of books written on the origin of the universe that do not really address the critical origin-related problems. Some of this fatigue stems from people not knowing which books are correct or will satisfy their needs. By customizing my books for different prospects, my theory allows everyone to focus on the products and services that fit their scientific and lay backgrounds, religious and nonreligious views, beliefs, and disbelief, regardless of age.

38. Undeniable reconciliation of science and the Biblical account of creation

Since the beginning of humanity, I am humbled to report that only I (Nathanael-Israel Israel), and nobody else, has offered a scientific explanation of the formation of the universe that perfectly matched science and the Biblical account of creation. Indeed, unlike anybody else before me, I am the first person in the whole world that calculated the mathematical equations that scientifically demonstrated that the Earth was formed 2.82 days, the Moon 3.32 days, and the Sun 3.69 days after the beginning of the universe, therefore confirming the 3500-year-old Biblical account of creation according to which the formation of the Earth was completed on the 3rd day. In contrast, that of the Moon and the Sun was completed on the 4th day of creation. I am also the first person in history to scientifically demonstrate that each day in the Biblical account of creation was literally 24 hours, a milestone that accurately reconciled science and the Bible, and that overturned a myth according to which some people have thought that each day of creation was millions of years (a misunderstanding that caused many people to deny God, the Creator). Therefore, I am honored to usher in a new era of proper understanding of the Biblical account of creation and its application to decoding the universe and its contents for the benefit of humankind.

Science180 Cosmology brings together Judeo-Christian believers through the accurate decoding and understanding of the Biblical account of creation, so they can better help humankind discover their true origin and the meaning of life. This theory catalyzes and convenes holistic expertise, advice, and influence on the crucial creationist issues facing believers worldwide, through customized programs and products that meet the needs of its clients and prospects. This discovery proved that of all the religious books in the world, only the Bible recounts a story of the formation of the universe that perfectly fits the scientific data, which science itself has failed to explain properly for centuries.

39. Use Science180 Turbulence 1 to test the intersection of science and faith accurately

My discovery is an opportunity for the secular world to recalibrate its methodology and get back to work. Some of my other books that can enlighten scientists quickly include:

- *Turbulent Origin of the Universe*
- *Turbulent Origin of Life*
- *Turbulent Origin of Chemical Particles*
- *Mathematical Proof of God's Existence at the Intersection of Science and Faith*
- *From Science to Bible's Conclusions*
- *Science180 Accurate Scientific Proof of God*
- *Reconciling Science and Creation Accurately*

Anyone who can do basic calculus can use a calculator to plug the scientific data from Table 1 into the formula I developed and test for themselves whether the Earth, the Moon, and the Sun were not formed within days, as the Bible says. For if you are still doubting my finding, what if your reasons for denying God's existence are wrong? Assume that what I said about the 4 simple variables is true. Then, even with those 4 assumptions, my theory still becomes reality. Although the Big Bang theory and evolution are filled with countless assumptions, some people may still prefer to embrace those theories over mine, which is supported by axioms demonstrated over centuries. I hope you won't follow that path. What if you admit your mistake and get some help before it's too late?

My discovery can help scientists refine their research methods better to understand the formation of the universe and even life. It can also help believers of all kinds test their faith more effectively. Therefore, this discovery offers an opportunity for believers and unbelievers worldwide to explore the intersection of science and faith without taking sides between reason and religion, and to know who is the God they must embrace so they don't believe in vain. As you consult my books *Turbulent Origin of the Universe*, *Turbulent Origin of Life*, and *Turbulent Origin of Chemical Particles*, you will also understand my theories about the origin of all forms of life and matter in the universe.

Whether scientists like it or not, my discovery will trigger a massive paradigm shift that must move modern science away from evolution and Big Bang theory to Science180 Cosmology. I know that many will come out and recklessly invent new theories as they compete against one another, even trying to compete against the truth. I am not interested in this kind of competition. I am not seeking applause from humans, and I want to be sure that, when all is done, I can hear from the one who sent me: "Well done!"

I have been telling the scientists and nonscientists that I discovered something huge, but some of them are not listening yet. Are they going to keep sleeping until others wake up and use my discovery against them? If you know anyone who can help relate this message to your leaders, please do so, so more people can partake in this paradigm shift. Whether they like it or not, secular research will never be the same again. Evolution and the Big Bang theory are DEAD, and they must die so that lives can be saved! Those who deny this fact and continue their research based on an evolutionary and Big Bang mindset will be shocked at how far behind they will be and crushed by their competitors, who will take the lead! For the truth is not in the path that evolution, the Big Bang theory, and all other billions-of-years theories put modern science!

Do you want to get out of the gigantic edifice built by evolution and Big Bang theory before it crashes, or do you want to stay in it at your own peril? For the scientists who may still doubt my finding, what if your reasons for denying God's existence are wrong?

Considering everything, why don't all creationists embrace Science180 Cosmology that debunked all of them scientifically, so they can hold hands together and help freethinkers discover the God they hate but who loves them? Why not embrace Science180 Creationism, which differs from all existing creationist theories, including Young Earth creationism?

TAKE-HOME MESSAGE OF SECTION 6: SCIENCE180 DIFFERENTIATORS

Can any scientific theory accurately prove the existence of God? The answer to that question is YES. And I just proved it to you with a theory (Science180 Creationism or Science180 Cosmology) that is:

- Easy-to-understand universe-origin and God's existence theory
- The state-of-the-art decoding experience of the origin of the universe
- The most accurate, reliable, safest, and best explanation of the universe's origin ever
- The universe-origin theory that helps you fight wasteful programs and theories
- The universe-origin rule breaker
- Bringing the Judeo-Christian believers together through the power of the accurate decoding and understanding of the Biblical account of creation, so they can better help humankind discover their true origin and the meaning of life
- The undeniable reconciliation of science and the Biblical account of creation

SECTION 6: SCIENCE180 DIFFERENTIATORS

- All the universe-origin solutions you love
- Capable of testing the falsification of the world's religions through the intersection of science and faith accurately.

To learn more, please visit www.Science180.com. Figure 26 is a visual element summarizing the breakthrough at the intersection of science and faith.

ACCURATE, SIMPLE, TRUSTWORTHY, NONCONFORMIST,
SCIENTIFICALLY VERIFIABLE PROOF OF
GOD'S EXISTENCE AT THE MATHEMATICAL
INTERSECTION OF SCIENCE AND FAITH
MATHEMATICAL INTERSECTION OF SCIENCE AND FAITH
SCIENCE AND FAITH CAN COEXIST, CHALLENGING THE NOTION THAT ONE MUST CHOOSE BETWEEN THEM.
BUT WHICH RELIGION/FAITH IS RIGHT?
T = D/Ve + 2Rπ/Vo
BIRTHDATE AFTER THE UNIVERSE'S BEGINNING
EARTH: 2.82 DAYS; MOON: 3.32 DAYS, SUN: 3.69 DAYS
LEARN MORE:
WWW.SCIENCE180.COM
WWW.ISRAEL120.COM
COPYRIGHT © 2025 BY DR. NATHANAEL-ISRAEL ISRAEL
Science180
2025
RELIGION / FAITH
RELIGION FIGHTING SECULAR SCIENCE
E=mc²
F=ma
a²+b²=c²
• EVOLUTION THEORY
• BIG BANG THEORY
• SCIENCE AT WAR WITH RELIGION
SCIENCE / RATIONALITY / REASON

Section 7: Conclusive Remark About my Argument for the Existence of God

"The real reason scientists have been struggling to accurately understand the universe-formation is because they have spent centuries collecting expensive, complicated, and massive amounts of data, but learned very little, if not nothing, about how to unconventionally step back to properly analyze it to decode the universe. Consequently, people learned to collect all kinds of data everywhere to build models and imaginary concepts that betray their discernment, but they never learned to unlearn wrong theories, nor learned how to stop trashing great raw data hidden in theories they dislike or misunderstand, never knew where to find and how to properly combine the fundamental variables without which it is impossible to ever clear the way so their data can properly work for and precisely lead them to the real origin of the universe. … To discover the intersection of science and faith, scrutinize the formation of the universe impartially, bravely, through the lens of Science180." — Nathanael-Israel Israel

"You are today where your thoughts have brought you; you will be tomorrow where your thoughts take you." — James Allen

"There is no conflict between science and the Bible. The same God who wrote the Bible is the God who made the world and all the worlds in the universe." — John MacArthur

"Our scientific power has outrun our spiritual power. We have guided missiles and misguided men." — Martin Luther King Jr.

"The more I study nature, the more I stand amazed at the work of the Creator." — Louis Pasteur

40. The "God's Existence Scientific Formula" that everybody is talking about

"In the beginning, God created the heavens and the earth." — Genesis 1:1

"In the beginning was the Word, and the Word was with God, and the Word was God. The same was in the beginning with God. All things were made by him; and without him was not any thing made that was made." — Gospel of John 1:1-3

Throughout the ages, scientists have discovered many formulas and given them different names. Science enabled scientists to collect massive amounts of raw data, but, in their ignorance, they misinterpreted the data, robbing God of the glory that is due to Him. Therefore, I would like to name the generic formula that calculates the birthdates of the celestial bodies the "**God's Existence Scientific Formula**." When understood and applied to the Earth, the Moon, and the Sun, this formula is sufficient to scientifically prove creation and the existence of God.

$$T = \frac{D}{Ve} + \frac{2R\pi}{Vo}$$

(Equation 42)

Where:

- **D is the Nathanael-Israel Israel Linear Distance, which for secondary celestial bodies is the distance between them and their primary body**
- **Ve is the Nathanael-Israel Israel Linear Speed, which for secondary celestial bodies is known as the escape velocity of their primary body**
- **R is the radius of the body**
- **Vo is the Nathanael-Israel Israel Gathering Speed, which for secondary celestial bodies is known as their orbital speed**

To glorify God who gave us His Word through the Bible (without which I could not have cracked the code of the turbulent origin of the universe), from now forth, the values representing the "Nathanael-Israel Israel Birthdate" of the Earth, Moon, and Sun will be called the "**Scientific Biblical Inerrancy Coefficients**":

- **2.82 days**
- **3.32 days and**
- **3.69 days.**

In other words, the "**Scientific Biblical Inerrancy Coefficients**" are 3 mathematical numbers demonstrated by Dr. Nathanael-Israel Israel, using an unconventional turbulence breakthrough, which, together, scientifically proved the inerrancy of the Biblical account of the creation of the celestial bodies. Understand the history of these numbers, and you will never misinterpret the origin of the universe again. Ignore them, and you will continue taking fictitious scientific and religious theories as true, therefore fooling yourself and your audience willingly! He who has ears, let him hear!

41. Does God exist? What is coming up next after my argument for the existence of God

Because science proved the "**Scientific Biblical Inerrancy Coefficients**", meaning the three cosmic timelines coded in the Biblical account of creation, revealed about 3500 years ago, long before science was born, we need to take seriously who the Bible said is the author: God. Moses did not just tell us the accurate timeline of the creation of the Earth, the Moon, and the Sun, but he also stated clearly the Creator. Moses said that the God of Israel is also the father of the spirit, spirits that physics will never prove using conventional tools. Since science confirms what the Bible said God did (see the Biblical account of creation in Genesis and also my summary in Figures 22, 24, and 25), we must also accept that God really exists and is expecting us to access Him through faith, which is the demonstration and the proof of invisible things that our current physical means cannot fully measure.

Formulated in 2025 by myself, Nathanael-Israel Israel, the 12 universal equations of God's existence are a system of 12 equations or questions based on 3 celestial bodies and 4 variables, which solutions led to the "**God's Existence Scientific Formula**" at the intersection of science and faith, therefore demystifying the so-called war between reason and faith, or between science and the Bible, or between rationality and religion. However, it is important to underscore that, science did not just prove that all faiths are rational, but my demonstration proved which faith is correct: the Judeo-Christian faith as explained in the Bible and practiced by Messianic Jews and Christians who believe in the literal interpretation of the Biblical account of creation, and who know that the Messiah has come and is returning back very soon to the Earth He created. In other words, when you consider the difference in belief between the non-Messianic Jews (i.e., those who reject Jesus Christ as the Messiah) and the Messianic believers (whether Jewish or Gentile), you can also know that not all who confess the Judeo-Christian faith are even serving the true God. I must boldly say that, Jesus Christ, known as Yeshua HaMashiach, is the true Messiah, which can lead us to God the Father! I detailed this mystery in my books *Reconciling Science and Creation Accurately* and *Science180 Accurate Scientific Proof of God.* The scriptures were revealed in such a way that, if we pay attention to the details, we can scientifically determine which religion is correct. I have sacrificed 12 years (2013-2025) after my PhD in the USA to help humankind discover its creator scientifically. I cannot arrive at this point of my writing without pointing you to that God, even if you may deny Him. At least, I know I accomplished my mission. Otherwise, it would be a big mistake to make you believe that all religions lead to God. For the Bible says that Jesus is the only way!

I am honored and humbled to have opened a new way for scientists and nonscientists to think about the origin of the universe, paving the way for greater discoveries and innovations for years to come. I know these discoveries honor God and will help humankind tremendously. Consult my book *Science180 Accurate Scientific Proof of God* to see more of the things that will happened next.

Considering the impact of my discovery, although Christianity has declined at lot in the last decades due to many reasons including the rise of Islam, fake news (from all political parties and ideologies nor just Democratic or Republican or Independent only), the negative impact of evolution and Big Bang theory on students and parents (who are trained or who train themselves to deny God, just as some believers do without anyone else influencing them), I predict that the trending hate toward God will and must change now. The world is going to see the unbelievable rise of Christianity and the Messianic movement across the globe like never before, with the demonstration of the power of God, not just through religion, but even through science, as all religions that cannot prove their God scientifically will fall, until the end of this world, that I cannot talk about here yet, comes! But as you stay tuned, you will be informed!

The world is full of so many religious and complicated theories, of course, but people don't know they are believing in the wrong things. Each existing creationist theory has its own baggage, which we can no longer tolerate, considering the light that Science180 creationism shone on the scientific evidence that was forced to cause people to deny God scientifically. Ignorance can no longer be excused! I did not just demonstrate the existence of God to you. Still, I must also emphasize that my discovery (which I was instructed to publish in 2025) is a wake-up call for the whole world and the beginning of a new era that most people won't believe we have entered since then! I wish you could listen to me so I could talk to you more about these mysteries you're starting to understand scientifically! But because some people don't know that they don't know, they don't want to seek the true knowledge that can free them. Now that you know there is a lot you ignore, I hope you will commit your life to knowing better, so that, slowly, you can finally seek and find what you don't know yet. Those who will have the opportunity to join me at Science180 Academy (www.Science180Academy.com) will be better enlightened about things I cannot convey in my books.

Considering everything, the most accurate theory about the formation of the universe is **Science180 Cosmology**, whose religious aspects I encapsulated in **Science180 Creationism**, regardless of any religious or philosophical version. What is it? See my book on that topic.

TAKE-HOME MESSAGE OF SECTION 7:

I spearheaded the "**God's Existence Scientific Formula**" that proved that:

- the Earth was formed in 67.66 hours = 2.82 days (meaning on the 3rd day of creation);
- the Moon in 79.79 hours = 3.32 days (meaning on the 4th day of creation);
- the Sun in 88.63 hours = 3.693 days (i.e., the 4th day of creation)

Therefore, I confirmed the existence of God, who revealed to Moses in the Biblical Book of Genesis that he is the Creator of the universe. The generic version of **God's Existence Scientific Formul**a is:

$$T = \frac{D}{Ve} + \frac{2R\pi}{Vo}$$

Where:

- **D is the Nathanael-Israel Israel Linear Distance, which for secondary celestial bodies is the distance between them and their primary body**
- **Ve is the Nathanael-Israel Israel Linear Speed, which for celestial bodies is known as the escape velocity of their primary body**
- **R is the radius of the body**
- **Vo is the Nathanael-Israel Israel Gathering Speed, which for celestial bodies is known as their orbital speed**

To glorify God, I coined the term "**Scientific Biblical Inerrancy Coefficients**" to designate the "Nathanael-Israel Israel Birthdate" of the Earth, Moon, and Sun: **2.82, 3.32, and 3.69 days, respectively.**

Name	Nathanael-Israel Israel Linear Distance (km)	Nathanael-Israel Israel Linear Speed (km/s)	Radius (km)	Nathanael-Israel Israel Gathering Speed (km/s)	Nathanael-Israel Israel Linear Time (Hours)	Nathanael-Israel Israel Gathering Time (Hours)	Nathanael-Israel Israel Birthdate (Hours)	Nathanael-Israel Israel Birthdate (Days)
Sun			696,000	19.4		62.584	88.630	3.693
Mercury	57,910,000	617.6	2,439.7	47.36	26.046	0.090	26.136	1.089
Earth	149,600,000	617.6	6,378.1	29.78	67.286	0.374	67.659	2.819
Moon	384400 (with respect to the Earth)	11.186	1,738.1	1.023	9.55 (with respect to the Earth)	2.963	79.79 (with respect to the Sun)	3.32 (with respect to the Sun)

Figure 27: Key details of the formation of the Earth, the Moon, and the Sun illustrating their Nathanael-Israel Israel Linear Distance, Speed, and Time, their Nathanael-Israel Israel Gathering Distance, Speed, and Time, and their Nathanael-Israel Israel Birthdate

SECTION 7: CONCLUSIVE REMARK ABOUT MY ARGUMENT FOR THE EXISTENCE OF GOD

Figure 27 illustrates the 7 key ingredients to break the code of the origin of the universe:

- Nathanael-Israel Israel Linear Distance
- Nathanael-Israel Israel Linear Speed
- Nathanael-Israel Israel Linear Time
- Nathanael-Israel Israel Gathering Distance
- Nathanael-Israel Israel Gathering Speed
- Nathanael-Israel Israel Gathering Time
- Nathanael-Israel Israel Birthdate

He who can understand these scientific ingredients and the historic recipe I revealed in this book and others I wrote has the scientific code of the universe's formation with him; he who cannot or doesn't want to will stay behind.

Do you want to know why I am teaching the whole world how science accurately supports the Biblical faith … and what can you do to avoid the dangerous thing that will happen next? Do you want to know the twenty four questions you cannot ignore if you want to decode the leaked secret about what will occur next in the world, so you can win? Are you interested in my unprecedented warnings to the whole world that you cannot afford to ignore if you want to be happy and avoid the deadly mistakes most people make, so you can life a profitable life? If yes, get your copy of my book *Science180 Accurate Scientific Proof of God* to discover the answer.

You know what? I am very glad you made it to the end. Before I close this book, I sincerely thank you for taking the time and attention to read it. It says a lot about you. By being here, you have demonstrated that you really care about understanding a unique, unconventional, nonconformist approach to the origin of the universe and the scientific proof of God's existence that I unveiled in this book. You prove that you really care about scrutinizing the intersection of science and faith. I suggest that you get my other books to uncover other aspects of my demonstration, and please share the news with your friends and family. If you have any questions, please visit my personal website (www.Israel120.com), where you can easily contact me. Get also some of my other books and services at www.Science180.com. If you forget everything in this book, please remember that, using scientific data collected over thousands of years, and verified and published by NASA and other top space agencies across the globe, I, Nathanael-Israel Israel, have scientifically demonstrated that the Biblical account of creation is scientifically true; therefore, God is real, and we need to do what He expects from us. Until I hear from you next, stay watchful of the development to come. Thank you again.

2026-4-24

Useful Resource for You

"Change is inevitable. Growth is optional." — John Maxwell

"Without my Saviour I am nothing. The Lord has guided me. He has shown me the way, just as he will show everyone who turns to Him." — George Washington Carver

42. Science180 Overview

42.1. Science180 Academy

What is Science180 Academy and how to register for it? Science180 Academy is the only non-degree training in the world that provides accurate, condensed, and comprehensive teaching on the origins of the universe, life, and chemical particles and the mathematical proof/test of God's existence. Regardless of your background, you will learn how to bypass technical knowledge that restricts even the experts from accessing the bottom of scientifically locked origin-related secrets contained in the massive scientific data that top scientists failed to decode properly.

When people register for Science180 Academy, they choose the program they want to focus on so their training can be personalized appropriately. This is like how people register at a university to take classes in a specific department that matches their needs! Science180's breakthroughs are so complex and dense that it is not realistic to try to explain all of it in just one program, else people will be overwhelmed, disinterested, and confused. In other words, Science180 Academy offers a wide range of origin-related training, allowing you to choose the most suitable for your needs so that, regardless of your background, you can equip yourself and align your mindset with the accurate explanation of the origin of the universe, so that your life can be improved. To register for Science180 Academy, please visit Science180.com/Science180Academy

You can find answers to some key questions I know you already have at Science180 Media Kit www.Science180.com/mediakit. You can also visit my personal website at www.Israel120.com

42.2. Top questions that Nathanael-Israel Israel can answer

What are some unique talk topics that I, and nobody else, can offer accurately, scientifically, on this Planet-Earth without upsetting believers and non-believers?

For many centuries, most scientists and laypeople have believed it was impossible to demonstrate the intersection of science and faith mathematically. I proved that this myth is false. Specializing in critical questions at the crossroads of science and faith, or rationality and religion, if you prefer, I have addressed more key topics at the intersection of science and faith than anyone else on Earth.

For a complete list of these topics, please visit www.Science180.com/topquestions or scan the following barcode:

I am available for media interviews, keynote speeches, seminars, workshops, consulting, and virtual trainings, aiming to help people of all ages and backgrounds understand the origins of the universe, life, and chemicals through a rational, impartial, and scientifically accurate lens. My mission is to help you decode complex questions about origins without taking sides between reason and faith, promoting a balanced understanding at the intersection of science and faith.

42.3. Other Science180 important resources

I can give Business Keynotes, Seminars, Workshops, Face-to-face & Virtual Trainings on any topic related to his areas of expertise. To hire me or Science180 to train, mentor, or give you a one-on-one consulting visit:

- Science180 Academy: Science180.com/Science180Academy
- Science180 Seminars: Science180.com/seminars
- Science180 Interview: Science180.com/interview
- Speaking Engagement: Science180.com/speaking
- Science180 Publishing: Science180.com/publishing
- Science180 Consulting: Science180.com/consulting
- Science180 Masterclass: Science180.com/masterclass

- To publish your book with Science180, the American publishing company that published the groundbreaking discovery about the origin of the universe, of life, and of chemicals, go to Science180.com/publishing
- Go to Science180.com/services to explore other ways Nathanael-Israel Israel can help you
- To register for the Science180 Interview Report, the only newsletter to read for guests and unconventional show ideas at the intersection of science and faith, with emphasis on the origin of the universe, the origin of life, the origin of chemical particles, and the scientific proof of God's existence, please visit Science180.com/registrationinterviewreport
- Subscribe to Science180 Newsletter: The only accurate universe-origin, life-origin, and chemicals-origin newsletter in the whole world! Science180.com/newsletter
- Collaborate or partner with Nathanael-Israel Israel: Science180.com/collaboration
- Ask for Review: Science180.com/askforreview
- Science180 Feedback: Science180.com/feedback
- Donate and support Nathanael-Israel Israel's efforts and initiatives: Science180.com/donate
- Quantity discounts: Purchase Science180 books, including this one, in bulk at a special discount: Science180.com/discount

USEFUL RESOURCE FOR YOU

- Recommend Nathanael-Israel Israel's books to your organization: Science180.com/books
- Share your feedback, criticism, testimony, experience, adventures, story, or comment about this book with me: Science180.com/feedback
- Share Nathanael-Israel Israel's groundbreaking discovery with others
- For all other inquiries, please go to Israel120.com/contact to contact Nathanael-Israel Israel today for more details.
- Follow Nathanael-Israel Israel on social media: Israel120.com

43. Other great books by Dr. Nathanael-Israel Israel

This book you are reading now is just a small tip of the gigantic iceberg representing the fantastic discovery I made concerning the origins of the universe, life, chemicals, and the existence of God. In fact, before this book, I published 9 others in 2025. Because I know that, depending on your background, some of them will greatly help you, I summarized below their content so you can better know which ones to get to continue your journey of properly decoding the origin of the universe and its content. To get these books please scan the next barcode of visit Science180.com/books.

43.1. Turbulent Origin of the Universe

There is Only One Scientific, Simple, Safe, Trustworthy, Unexpensive, Brave, Practical, Nonconformist, Universal, Verifiable Formula that Accurately Decodes the Universe Formation … But You Are Not Using It

THE FIRST AND ONLY SCIENTIFIC BOOK THAT ACCURATELY EXPLAINS EVERYTHING YOU NEED TO UNCONVENTIONALLY, EASILY, AFFORDABLY, AND ENJOYABLY DECODE THE UNIVERSE FORMATION

In *Turbulent Origin of the Universe,* filled with great diagrams and digestible scientific facts, you will discover, learn, or get:

- This all-in-one, proven, straightforward scientific formula precisely uncovers the formation of the universe and explains the birthdates of stars, planets, satellites, asteroids, and other celestial bodies. Use this knowledge to surpass your competitors and avoid costly mistakes that many others have made at their own risk.

- Extraordinary, unprecedented, accurate insights into the first factors (e.g., early universe physics) that defined the history and formation of the universe, so you can tap into deep scientific secrets you ignore, and set yourself apart from others
- The new physics that will revolutionize science forever and land you in a zone of original ideas that improve lives nonstop, regardless of your expertise
- The four simple principles without which it is impossible for anyone to truly understand the formation of the universe, think clearly, work differently, achieve goals, or perform at a higher level for better results.
- The verified key to move the cosmological mountains of misunderstanding, so you can confidently free your mind from doubts, improve your health, and prevent you from any danger connected with sticking with wrong assumptions
- Save time and money, and enjoy your life once you remove errors holding your true understanding of the universe-origin captive
- The scientific formula that successfully tested the existence of God in a way that shocked believers, skeptics, and all other freethinkers
- Why the scientific community has failed to sufficiently explain the origin of the universe, and understand how existing theories have missed and imposed limits on the vision of scientists
- Gain specialized knowledge, current information, and innovative ideas to grow your market, reduce unnecessary expenses, avoid wasting time on ineffective projects, and concentrate on profitable solutions
- How Science180 Academy can strategically enlighten you, guide you to navigate and filter the extensive data collected about the universe, enabling you to address the world's most challenging questions, eliminate any scientific and philosophical obstructions that may hinder you, and bring you significantly closer to achieving your best life.
- How to better resonate with your target market that is craving something original that breaks the wrong explanations of the universe-origin

Get *Turbulent Origin of the Universe* today to begin an incredible journey of accurately decoding the universe and changing your life forever!

Learn more at (Science180.com/scientific).

43.2. Turbulent Origin of Chemical Particles

Why You Don't Have to Embrace Evolution, Big Bang, or Deny God to Scientifically Prove the Formation of All Chemical Particles

USEFUL RESOURCE FOR YOU

FIND ALL THE RELIABLE, CONVINCING, SCIENTIFIC ANSWERS YOU NEED TO SUCCESSFULLY DECODE THE ORIGIN OF CHEMICAL PARTICLES SAFELY

Where did all elementary and composite particles, such as atoms, molecules, minerals, and rocks, originate? What are the core factors, mechanisms, and universal processes that determined their formation and characteristics? What were the nature of their precursors at the universe's beginning, and what fundamental processes shaped or formed them into the chemicals we observe today? What drives the abundance and variety of chemicals in celestial bodies across the universe? How are the formation of chemical particles connected to the creation of galaxies, stars, planets, asteroids, and satellites? What insights about chemicals can help explain the true cause and significance of gravity and other fundamental forces? How does the development of chemical particles fit into the overall story of the universe's formation?

After studying these questions for over twelve years, Dr. Nathanael-Israel Israel has discerned that understanding the origin of chemical particles is a highly challenging yet rewarding endeavor that necessitates original, scientific, mathematical, and philosophical efforts beyond the current state of modern science, until recently. The resolution to these complex issues is found in *Turbulent Origin of Chemical Particles*, a clear and reliable publication designed to facilitate rapid, cost-effective, and efficient comprehension of all essential knowledge needed to resolve the fundamental questions regarding the origin, formation, and function of chemical particles. Whether you are a chemist, a biochemist, or a scientist or engineer from any discipline, provided you possess a reasonable background in chemistry but lack a scientific understanding of the origin of chemical particles, this remarkable book is indispensable. Exceptionally rich in insightful analysis, with impressive graphs, tables, and the historic formula that unlocked the universe's origin, *Turbulent Origin of Chemical Particles* will:

- Make it easier than ever for you to properly understand, decrypt, and articulate the real origin of natural chemical particles in the universe, therefore freeing you from false and boring explanations of the origin of all matters, and embrace the proven theory that opens doors to unparalleled opportunities
- Professionally guide you on transforming your true knowledge of the origin of chemical particles into valuable insights quickly, helping you

establish yourself as a symbol of freedom, power, creativity, and originality in your field.

- Fire you up to become the best version of you, and to cause positive changes to your initiatives that will profit you nonstop
- Explore exciting illustrations and unique explanations of how all matter in the universe was formed, written in clear language that brings us closer to understanding the core mysteries of chemistry. This approach paves the way for future advancements in technology, innovation, discoveries, and breakthroughs.
- Equip you to bypass technical knowledge that restricts non-experts from accessing the origin-related secrets contained in the massive scientific data, and get to the bottom of origin-related mysteries regardless of your background, so you can empower yourself to leave unforgettable marks in your field of expertise. More at Science180.com/chemical

With *Turbulent Origin of Chemical Particles*, the accurate deciphering and understanding of chemical formation has never been profitable or easy. Hence, this great book is THE ultimate how-to guide for great people who want to correctly decode the origins of chemicals and positively transform their lives. Get this celebrated book today. Don't wait!

43.3. Turbulent Origin of Life

Why You Don't Have to Embrace Evolutionism or Check Your Brain at the Door in the Name of Faith or Science to Accurately Decrypt the Origin of Life Using the Historic Formula of the Universe Formation

THE ONLY ACCURATE FORMULA TO SCIENTIFICALLY EXPLAIN THE FORMATION OF ALL FORMS OF LIFE QUICKLY

Every human being will benefit from understanding the real origin of life. But the problem is that most efforts to explain the origin of life are complex, inaccurate, confusing, partisan, and complicated, thereby creating serious challenges for those eager to scientifically decrypt the origins of all forms of life. Most people want an accurate, simple, straightforward, nonpartisan life-origin book that is free of jargon and difficult concepts known only to experts. This elegant scientific book breaks down the technicality of the origin of life in a language that even nonscientists can easily comprehend. It is a trustworthy book that will help you to quickly, cheaply, easily, and efficiently navigate everything you need to know to finally decode and solve the puzzling problems about the origin of life, while also giving you a crash course on the origin of the universe.

USEFUL RESOURCE FOR YOU

Unlike any book you have ever read on the origin of life, this historic masterpiece (which distills complex scientific data into simple explanations that make sense) is the starting point for any smart person wanting to rationally understand the formation of all living things. By the time you finish reading *Turbulent Origin of Life*. You will discover:

- Why, in spite of the massive amount of scientific data collected on living things, have scientists misunderstood the formation of life until now, and then uncover in a simple language the one thing that was needed to accurately crack the code of life, but that scientists have missed and that has been causing them headaches, overwhelm, and burnout
- Step-by-step pathway to decode the origin of life and get the power, freedom, and boldness to take advantage of the opportunities that an accurate understanding of the origin of life creates
- The high connection between the code of the universe formation and the process by which life on Earth was formed, so you can become a fulfilled thought leader in your field of expertise
- Tools to stand as a lightning bolt that electrifies those who are still struggling to understand the formation of all forms of life in the universe
- Strategies to expand human understanding of the mysterious, supernatural, and impossible that hinder progress.
- Scientific approach to holistically detect, correct, and remove all misinformation, ambiguity, and misleading claims and theories surrounding the origin of life

Whether you are a scientist or a layperson, a believer or a skeptic, it is imperative not to overlook the more advanced, improved, swift, straightforward, cost-effective, convenient, and precise formula unveiled in this significant publication that has triumphantly deciphered the origin of life. Get *Turbulent Origin of Life* today and make a transformative impact! Do not delay. More at Science180.com/life. Connect with Dr. Israel at Israel120.com.

43.4. From Science to Bible's Conclusions

How Decoding the Universe-Origin by Properly Revisiting Scientific Data—That Top Scientists Collected but Wrongly Analyzed—Bizarrely led to the 3500 Years Old Biblical Account of Creation

BOLDEST SCIENTIFIC FORMULA OF GOD AND CREATION

THE # 1 UNIVERSE-ORIGIN MASTERPIECE OF ALL TIME ... AND THE MOST ACCURATE SCIENTIFIC FORMULA THAT STOOD AND WILL STAND THE TEST OF TIME AND OF MATHEMATICS

The real reason scientists have been struggling to accurately understand the universe formation is that they have spent centuries collecting expensive, complicated, and massive amounts of data, but learned very little, if anything, about how to step back unconventionally to properly analyze it to decode the universe. As a result, people have learned to gather all kinds of data everywhere to create models and imaginary ideas that betray their judgment. However, they have never learned to unlearn false theories or how to stop discarding valuable raw data hidden in theories they dislike or misunderstand. They also lack knowledge of where to find and how to properly combine fundamental variables, which are essential for clearing the way so that their data can effectively work for them and lead them to the true origin of the universe. The question remains: how can people abandon dangerous theories they believe are correct when they don't know any better alternatives?

Lucky you, that is where Dr. Nathanael-Israel Israel, the founder of Science180 (Science180.com), came in to properly reanalyze and put under control these costly, underrated data to provide the accurate, simple solution people have been looking for throughout the ages, but have ignored. In *From Science to Bible's Conclusions*, you will:

- Get a world-class explanation of the 4 fundamental variables without which it is unquestionably impossible to ever decode the universe-formation scientifically
- Save time and money, and enjoy a life filled with the wonderful peace that the accurate understanding of the universe-origin can create
- Discover the errors in the scientific theories and religious belief systems about the universe-formation that are putting you at risk, and learn how to take control over cosmological threats lurking at the edge of your rational mind, faith, disbelief, or doubt
- Unlock the accurate scientific formula to rationally test the existence of God in a historic way that uncompromisingly satisfies both believers and skeptics (Science180.com/public)
- Get all you need to become a knowledgeable person who will never again need anybody else to explain to you the origin of the universe, for you will fully understand and articulate it yourself, and rationally know whether science is really at war with religion
- Receive deep insights that even those who went to university for years were unable to decrypt, so you can equip yourself to eliminate all forms of universe-origin prejudices
- Discover whether the scientific data finally confirms that the formation of the Earth was completed on the 3rd day, while that of the Moon and the

Sun was on the 4th day of creation, like the Bible says, or whether the data proves that it took billions of years to progressively form the universe

- Understand the celebrated scientific formula that rationally puts to rest all debates about the relationship between science, faith, and all theories about the universe-origin so you can properly develop yourself, expand your network, and shape your future

Take a moment to explore this fascinating, scientifically verified bestselling book that offers the clear, awe-inspiring answers you've been curious about. It's perfect for anyone—believers, skeptics, or freethinkers—seeking to understand the universe better. Don't miss out!

Dr. Nathanael-Israel Israel is a leading expert in understanding the origins of the universe, life, and chemicals. He founded Science180 Academy (www.Science180Academy.com). Connect with this inspiring scientist and discover free resources by visiting Israel120.com.

43.5. Reconciling Science and Creation Accurately

What Science Accurately Teaches about Creation and God's Existence that Atheists, Freethinkers, and even Most Christians Ignore … And How to Demonstrate it Without Taking Sides Between Rationality and Faith

THERE IS ONLY ONE SIMPLE, COMPELLING, SOLUTION-DIRECTED SCIENTIFIC FORMULA ACCURATE ENOUGH TO RATIONALLY EXPLAIN HOW GOD CREATED THE UNIVERSE

Reconciling Science and Creation Accurately is a landmark book offers a unique perspective on universe-origin topics, authored by one of today's most esteemed thinkers. It investigates, from a scientific standpoint, the most profound questions that captivate believers, skeptics, and free thinkers alike: How can one logically verify, without abandoning reason in the name of faith, that God created the universe? What are the origins of the universe, and through what mechanisms did God bring it into existence? Do these mechanisms still function today? Can believers dismiss incorrect theories if they believe science cannot prove the Genesis story definitively, or if they see science as opposed to faith? Moreover, how can believers effectively demonstrate the Biblical account of creation to skeptics through scientific reasoning?

Lucky you, Dr. Nathanael-Israel Israel successfully navigated all those questions with an accuracy that has been applauded by both scientists and nonscientists across the globe. After reading *Reconciling Science and Creation Accurately*, you will confidently:

- Scientifically prove the Biblical account of the creation of the universe and God's existence in a way that makes the head of those who deny God spin faster than a DJ's turntable
- Know how to rationally talk to anti-creationists, evolutionists, Big Bang proponents, atheists, skeptics, and other freethinkers about the universe formation, and they will beg you to know more about God, the Creator, whom they mistakenly rejected
- Discover very accurate, rare, and factual truths about the universe-origin that will save you time and money, and get you much closer to the better and joyful life you want to live today and forever
- Improve your health and faith by knowing that the existence of God can be scientifically justified using Science180 Cosmology and particularly Science180 Creationism
- Enter a new area of freedom and power by crushing the head of and breaking free from the suffocating expectations of all wrong theories that have hijacked secular and religious education, and that have held Biblical creation captive for 3500 years
- Break free from the suffocating expectations of some forms of creationism that have sequestered the minds of some believers for a long time
- Uncompromisingly, intelligently, and scientifically explode the myth of those who, instead of literally taking the Biblical days of creation as 24-hour consecutive days, think that they were millions of years, or were representative of long ages, or that millions of years existed before them or were positioned between them
- Understand the accurate standard to interpret the Biblical account of creation thanks to Science180's breakthrough that transformed science and laid a foundational bedrock for the inerrancy of Scripture. More at Science180.com/biblical

Now that Genesis—the oldest manuscript in the world, predating science and most religions—is confirmed scientifically accurate, what profound and unstoppable paradigm shift will occur when science and the Bible align perfectly? This insightful book explores what happened at the beginning, what lies ahead, and why it's crucial to rethink everything you've been told about the universe's origin now—so you won't regret it later. Don't say nobody warned you!

Dr. Nathanael-Israel Israel is known as the first person in the world to discover the historic all-in-one, proven, simple, scientific, mathematical equations that scientifically demonstrated that the Earth was formed 2.82 days, the Moon 3.32 days, and the Sun 3.69 days after the beginning of the universe, therefore using scientific data collected by top scientists throughout centuries to confirm the Biblical account of creation that revealed about 3500 years ago that the formation of the Earth was completed on the 3rd day, while that of the Moon and the Sun was completed on the 4th day of creation. More at Israel120.com.

43.6. How God Created Baby Universe

What Children Must Scientifically Learn Early about the Universe Formation to Avoid Dangerously Abandoning God Later in Life Just Like Most College Students Who Embrace Evolution and Big Bang That Deny Biblical Creation

THE FIRST AND ONLY BOOK THAT ACCURATELY EXPLAINS EVERYTHING ABOUT THE FORMATION OF THE UNIVERSE AND LIFE IN A WONDERFUL LANGUAGE THAT ALL CHILDREN AGES 7-12 CAN EASILY, FULLY UNDERSTAND & ENJOY!

As the only universe-origin book that your entire family can enjoy together, *How God Created Baby Universe* helps children understand the universe's formation early in life. It also teaches them how to identify errors in theories or stories that might mislead them as they grow up. Adding this valuable, reliable, and affordable book to your child's educational journey is a strategic step toward their bright future.

With *How God Created Baby Universe*, you will:

- Have peace of mind that children will get accurate, fit, and easy-to-understand universe-origin information that will produce real results in their lives
- Become the leader that captures the heart of children craving for the original explanation of the formation of the universe, so you can clear their way for freedom, power, technology, innovation, and breakthroughs of the future (learn more at Science180.com/children)
- Protect yourself and loved ones from misleading theories in literature and media by ensuring children are safe and empowered with accurate knowledge of the universe's origins.
- Explain complicated secrets to children about how to locate mistakes in origin-related theories, so you can save time, money, and other resources to improve their lives
- Ultimately, boost children's confidence in detecting, confronting, and avoiding wrong theories by knowing the facts and real processes involved in the formation of the universe
- Help children to easily sort out their origin-related questions using strategies that get them to tap into deep secrets that even highly educated people ignore

- Clearly explain to children how to mathematically know without a doubt whether God created the universe as the Bible says, or billions of years evolution processes formed it

Explaining how the universe and life formed to children is challenging, but with *How God Created Baby Universe*, you'll discover a proven approach to help children confidently and joyfully understand these big questions. They will likely laugh and appreciate this book, making it a great purchase for the children in your life today.

Dr. Nathanael-Israel Israel, a member of the American Association for the Advancement of Science. More information is available at Israel120.com.

43.7. How Baby Universe Was Born

How to Scientifically Talk to Children about the Universe Formation and They will Know Forever How to Correctly Test the Intersection of Science and Faith

THE ONLY SCIENTIFIC BOOK THAT CORRECTLY DEMONSTRATES THE FORMATION OF THE UNIVERSE TO CHILDREN AGES 7-12 WHO OR WHOSE PARENTS DON'T BELIEVE IN GOD, AND THEY WILL PROPERLY, SCIENTIFICALLY KNOW FOR SURE WHETHER GOD CREATED THE UNIVERSE OR NOT

In our modern, secular world, parents face many worries about what children learn at school and elsewhere, over which they have little control once kids leave home. However, this isn't inevitable. Dr. Nathanael-Israel Israel, a scientist specializing in universe and life origins, emphasizes the importance of parents equipping their children with a strong worldview that balances science and reality. This helps prevent children from being swayed by secular education or religious beliefs. With *How Baby Universe was Born*, you'll learn how to:

- rationally explain the origin of the universe to children in our modern world
- use the latest universe-origin breakthrough to scientifically educate children so they don't become victims of wrong theories and dangerous dogmas that, in the name of reason or faith, try to enslave their understanding of the real origin of the universe and of life
- scientifically help children understand why wrong science is not the only danger of trying to understand the universe's origin, but wrong belief as well

- scientifically help children to test for themselves whether science and faith can meet, so they become the best version of themselves through the correct understanding of the origin of the universe
- if you don't believe in, dislike, or doubt the existence of God, but want your children to learn about the universe's scientifically-supported origins, this book is for them. Get *How Baby Universe Was Born* and start reading today. For more information before purchasing, visit Science180.com/children.

Dr. Nathanael-Israel Israel is a Beninese-American scientist, who passionately shows people of all ages and backgrounds how to scientifically understand the formation of the universe, life, and chemicals. To discover more about him, visit Israel120.com.

43.8. Origin of the Spiritual World

Top Secrets about the Origin of Everything in the Universe that Some Elites Have Hidden from You for Thousands of Years

ONLY ONE ANCIENT BLUEPRINT HAS THE RELIABLE POWER TO HELP YOU TO ACCURATELY DECRYPT THE SPIRITUAL ORIGIN AND HISTORY OF EVERYTHING IN THE UNIVERSE

Many books discuss the origin of the universe and life, but this remarkable book is the first and only one to undeniably explain how the formation of the universe and everything within it was genuinely revealed in secret and rejected scriptures, such as the Books of Enoch and others. In *Origin of the Spiritual World*, you will:

- Discover deep, rejected secrets that have prevented humankind from unearthing the beginning of the universe
- Plainly see the scientific evidence, concealed within scriptures, clearly illustrates the formation of the Earth, the Moon, and the Sun within a matter of days. This historic revelation astonishingly aligns with scientifically verified data, as documented in the publication *From Science to Bible's Conclusions* authored by Dr. Nathanael-Israel Israel.
- Use the lost and rejected scriptures effectively to explain how the universe was created. This understanding can enhance your knowledge of the Bible, inspire innovation in your field, and contribute to your continuous self-improvement.
- Empower and align yourself with the historic breakthrough that has done what no other discovery has ever done: accurately unlock and decode

mysteries concerning the origin of the cosmos and its content using scientific keys revealed in ancient scriptures that some elites have concealed (Science180.com/pseudepigraphic)

- Discover and apprehend the complex formation of the universe and life without leaving out the challenging questions that people of all ages have been struggling to answer for thousands of years, while the answers were hidden
- Discover greater joy in life through a clear understanding of both ancient and contemporary revelations regarding the creation of the universe, remarkably supported by modern science, which some individuals mistakenly believe contradicts the Bible.
- Make a difference and blaze new trails for those who depend on your leadership

If you believe in God, have some origin-related questions that you cannot find answers to anywhere, not even in the Bible, and if you want to tap into historically neglected revelations to answer fundamental universe and life questions, then be sure to get a copy of *Origin of the Spiritual World* today.

Dr. Nathanael-Israel Israel is the author of *Turbulent Origin of the Universe, Reconciling Science and Creation Accurately, Turbulent Origin of Chemical Particles, From Science to Bible's Conclusions, Turbulent Origin of Life, How Baby Universe Was Born, Science180 Accurate Scientific Proof of God.* Learn more about this rare scientist at Israel120.com.

43.9. Science180 Accurate Scientific Proof of God

Can We Scientifically Explain the Formation of the Universe Through Natural Processes Without Evoking Evolution and Big Bang?

THE FIRST AND THE ONLY SCIENTIFIC BOOK THAT TALKS TO ANTI-CREATIONISTS, EVOLUTIONISTS, BIG BANG PROPONENTS, ATHEISTS, AND ALL OTHER FREETHINKERS AND RATIONALISTS ABOUT THE UNIVERSE FORMATION, AND THEY BEG TO KNOW MORE ABOUT GOD, THE CREATOR, THAT THEY DENY.

As you read this historic book, you will:

- Know the one clear sign you should always watch for when trying to identify the primary cause and key drivers behind the fundamental processes that lead to the formation of the universe.

- Discover the only way to explore the question of God's existence scientifically and to find out which among the many beings worshipped worldwide might be the true God.
- Accurately answer the most critical universe-origin and life-origin questions so you can stop standing in tension with consequential question marks, including those related to religion and reason, or the so-called war between science and the Bible.
- Identify the errors in scientific and religious theories about the origins of the universe and life. Ignoring these errors risks your understanding and progress, so it's crucial to learn how to confidently and rationally address the threats that threaten your efforts to comprehend the universe and life today.
- Challenge the cosmological status quo and embrace the real change that will disrupt the hidden cages that may be holding you, and that you ignore
- Definitively answer all your doubts about the source or author of the universe and life … (learn more at Science180.com/godproof)
- Understand that religion or faith, alongside reason or science, can coexist and be properly harmonized to guide you toward the true source of everything in the universe.
- Satisfy your burning desire for freedom from beliefs and scientific theories about the universe-origin and life-origin that suffocate you and bind your mind, faith, unbelief, heart, and education
- Scientifically set on fire all false theories or dogmas about the existence of God, the Creator, that are enslaving humankind.

Whether you are a believer, unbeliever, freethinker, administrator, politician, curriculum designer, curriculum specialist, education policymaker, teacher, librarian, school board member, researcher, parent, student, clergy, or a layperson, as long as you are really seeking to scientifically understand the rational proof of the existence of God, *Science180 Accurate Scientific Proof of God* is the much-admired book written for great people just like you! Grab your copy today and start reading it! Don't wait any longer!

Dr. Nathanael-Israel Israel, #1 Universe-origin, Life-origin, and Chemicals-origin Expert. Connect with him and get free resources at Israel120.com.

43.10. Mathematical Proof of God's Existence at the Intersection of Science and Faith

THE HISTORIC, SCIENTIFICALLY VERIFIABLE THEORY THAT PROVED THE RATIONAL ABCS OF DECODING THE ORIGIN OF THE UNIVERSE USING JUST FOUR ESSENTIAL VARIABLES THEY DON'T WANT YOU TO KNOW ABOUT

BOLDEST SCIENTIFIC FORMULA OF GOD AND CREATION

This is the modern non-evolution, non-Big Bang theory that is scaring the scientific community today because it mathematically blows up a key foundation of modern science: the nonsensical or mythological aspects of evolution and the Big Bang. Here, you will discover the scientific framework that led to the three specific all-in-one formulas that cracked the code of turbulence (a long-standing unsolved problem in physics), and decoded the beginning of the universe and how modern scientific data, when analyzed through these formulas, confirms the Genesis timeline (formation of the universe in days) rather than a billions-of-years model. In this book, you will scientifically discover how the

- Earth was completed in 2.82 days (matching the 3rd day of creation)
- The Moon in 3.32 days (matching the 4th day) and
- The Sun in 3.69 days (matching the 4th day).

In *Mathematical Proof of God's Existence at the Intersection of Science and Faith*, you will also understand whether:

- A scientifically verifiable theory can destroy evolution and the Big Bang theory
- Any "dumb" scientist has ever won a Nobel Prize in Physics
- Any stupid scientific theory and religious dogma have really been ruling the world
- Mathematics can prevent science from arguing with some religions irrationally
- Math can help atheists to discover the God they hate
- Creationists are really explaining creation as they say
- The Genesis creation story can be proven as a scientifically verifiable history
- A Biblical argument can actually win a Nobel Prize in physics and math
- Scientists can rationally deal with their cosmological blind spot
- You can believe in God without faith, relying solely on scientific reasoning
- This theory is the proof of God's existence that rational scientists have been seeking
- There is rational proof of God's existence that all freethinkers will like

Get this amazing book today to answer all of those questions and many more, and start enjoying your life knowing for sure how the universe began! Learn more at Science180.com/MathematicalProof

Founder of Science180 (www.Science180.com), Dr. Nathanael-Israel Israel is the discoverer of the timeline of cosmic formation, which united science and faith while challenging traditional scientific timelines spanning billions of years. Learn more at www.Israel120.com.

44. Which of Nathanael-Israel Israel's books is best for me according to my background?

In 2025, I published nine books and in 2026, I published 2:

- *Turbulent Origin of the Universe*
- *Reconciling Science and Creation Accurately*
- *Turbulent Origin of Chemical Particles*
- *Origin of the Spiritual World*
- *From Science to Bible's Conclusions*
- *Turbulent Origin of Life*
- *How God Created Baby Universe*
- *How Baby Universe Was Born*
- *Science180 Accurate Scientific Proof of God*
- *Mathematical Proof of God's Existence at the Intersection of Science and Faith*
- *Boldest Scientific Formula of God and Creation*

To help you know which books can help you the best without wasting your time, please see below my selection according to your background.

I am an atheist. Which of Dr. Israel books do you recommend I read first?

Please start with:

- *Boldest Scientific Formula of God and Creation*
- *Science180 Accurate Scientific Proof of God*
- *From Science to Bible's Conclusions*
- *Mathematical Proof of God's Existence at the Intersection of Science and Faith*

Then, move to the others

I am a scientist. Which books do you recommend I read first?

- *From Science to Bible's Conclusions*
- *Boldest Scientific Formula of God and Creation*
- *Mathematical Proof of God's Existence at the Intersection of Science and Faith*
- *Science180 Accurate Scientific Proof of God*
- *Turbulent Origin of the Universe*
- *Turbulent Origin of Chemical Particles*
- *Turbulent Origin of Life*

I am a chemist, a life scientist. Which of Dr. Israel books do you recommend I read first?

- *From Science to Bible's Conclusions*

- *Turbulent Origin of Chemical Particles*
- *Turbulent Origin of Life*
- *Turbulent Origin of the Universe*
- *Science180 Accurate Scientific Proof of God*
- *Boldest Scientific Formula of God and Creation*
- *Mathematical Proof of God's Existence at the Intersection of Science and Faith*

I am a preacher. Which books do you recommend I read first?

- *Reconciling Science and Creation Accurately*
- *Boldest Scientific Formula of God and Creation*
- *Science180 Accurate Scientific Proof of God*
- *From Science to Bible's Conclusions*
- *How God Created Baby Universe*
- *How Baby Universe Was Born*

I am a scientist, mathematician, or philosopher wanting to drive real science from Dr. Nathanael-Israel Israel discovery. Which books should I read first?

- *From Science to Bible's Conclusions*
- *Boldest Scientific Formula of God and Creation*
- *Mathematical Proof of God's Existence at the Intersection of Science and Faith*
- *Turbulent Origin of the Universe*
- *Reconciling Science and Creation Accurately*
- *Turbulent Origin of Chemical Particles*
- *Turbulent Origin of Life*
- *Science180 Accurate Scientific Proof of God*

I am a parent wanting to help my children understand Dr. Nathanael-Israel Israel discovery. Which book do you recommend?

- *How Baby Universe Was Born*
- *How God Created Baby Universe*
- *Reconciling Science and Creation Accurately*
- *From Science to Bible's Conclusions*
- *Science180 Accurate Scientific Proof of God*
- *Boldest Scientific Formula of God and Creation*

To get a copy of Nathanael-Israel Israel's books that will help you benefit from his groundbreaking discoveries, go to Amazon.com or Science180.com/books. If you contact me at Science180.com, I can give you more than 45% discount if you order many copies.

References

George William K. (2013). An Introduction to Natural Convection Flows. Chalmers University of Technology Gothenburg, Sweden, 55 pages. Download from *www.turbulence-online.com* on October 19, 2017.

Islam for Today (2007). The Origin of Life: An Islamic perspective". Retrieved 2007-03-14. http://www.islamfortoday.com/emerick16.htm.

Israel Nathanael-Israel (2025a). Turbulent Origin of the Universe. Science180, Augusta, USA 683 pages.

Israel Nathanael-Israel (2025b). From Science to Bible's Conclusions. Science180, Augusta, USA 170 pages.

Israel Nathanael-Israel (2025c). Reconciling Science and Creation Accurately. Science180, Augusta, USA 299 pages.

Israel Nathanael-Israel (2025d). Turbulent Origin of Chemical Particles. Science180, Augusta, USA 397 pages.

Israel Nathanael-Israel (2025e). Turbulent Origin of Life. Science180, Augusta, USA 370 pages.

Israel Nathanael-Israel (2025f). Origin of the Spiritual World. Science180, Augusta, USA 151 pages.

Israel Nathanael-Israel (2025g). How Baby Universe Was Born. Science180, Augusta, USA 130 pages.

Israel Nathanael-Israel (2025h). How God Created Baby Universe. Science180, Augusta, USA 224 pages.

Israel Nathanael-Israel (2025i). Science180 Accurate Scientific Proof of God. Science180, Augusta, USA 214 pages.

Israel Nathanael-Israel (2026). Mathematical Proof of God's Existence at the Intersection of Science and Faith. Science180, Augusta, USA 74 pages.

NASA (2018). Planetary fact sheets. Fact sheets of the Sun, planets, satellites, rings and selected asteroids in the Solar System. Author/Curator: Dr. David R. Williams, NASA Goddard Space Flight Center, Greenbelt, MD, USA. http://nssdc.gsfc.nasa.gov/planetary/factsheet/. Visited on November 19, 2018.

Index

A

A. P. J. Abdul Kalam........................ 83
Abraham 3, 87, 90, 92, 98
Absurdity................................. 91, 96
Africa............................7, 13, 91, 162
Agnostics..........12, 86, 102, 103, 115
Albert Einstein ... 1, 12, 21, 22, 27, 32, 108, 109, 110, 111, 112, 113
Allah 92, 93
America 95
American..........3, 4, 6, 144, 145, 161
American Association for the Advancement of Science 6, 143, 144, 161
American Chemical Society...... 6, 161
American Society for Microbiology. 6, 161
American Society of Agricultural and Biological Engineers............ 6, 161
American Society of Agronomy 6, 161
American Society of Biochemistry and Molecular Biology................. 6
Ancestors ... 21, 54, 90, 91, 92, 93, 94, 110, 111
Angel Gabriel 92
Animism 91
Anti-creationists.............12, 142, 146
Arabs 90, 92, 93, 94, 95
Aristarchus.............................. 13, 14
Asteroids 136, 137, 151
Atheists 2, 12, 86, 97, 98, 101, 102, 103, 115, 142, 146, 148, 149
Atoms.. 137

B

Baby Universe 2, 98, 143, 144, 145, 146, 149, 150, 151, 161
Barack Obama 107
Beginning of the universe 126
Belief.............................. 91, 92, 111
Believers..... 2, 3, 8, 12, 85, 87, 89, 90, 96, 98, 101, 102, 106, 107, 108, 117, 120, 121, 127, 128, 133, 136, 139, 140, 141, 142, 147, 162
Ben Sweetland 96
Benin 7, 114
Beninese 67, 113, 145
Bible ... 2, 3, 87, 90, 91, 92, 93, 96, 97, 98, 100, 102, 105, 106, 108, 112, 115, 120, 121, 125, 126, 127, 139, 140, 141, 142, 143, 144, 145, 146, 147, 149, 150, 151, 161
Biblical.......................2, 139, 142, 161
Biblical account of creation. 2, 86, 87, 90, 92, 95, 99, 100, 102, 104, 105, 106, 111, 115, 117, 120, 122, 127, 131, 141, 142, 161
Biblical inerrancy......... 101, 126, 127
Biblical Inerrancy........................ 129
Big Bang theory.. 1, 2, 3, 4, 2, 3, 7, 14, 85, 98, 100, 104, 105, 106, 115, 117, 121, 122, 128, 136, 142, 143, 146, 148
Bill Gates 105
Billions of years 2, 3, 97, 98, 105, 110, 112, 122, 141, 144, 148
Biology 23, 161
Birthdate of celestial bodies ... 58, 73, 81
Buddhism 90, 91, 105

C

C. S. Lewis.................................. 103
Cascade of breakup....................... 24
Celestial bodies .. 4, 5, 6, 7, 11, 12, 13, 14, 15, 16, 17, 18, 19, 21, 22, 23, 24, 25, 26, 27, 28, 29, 30, 31, 32,

33, 34, 37, 38, 39, 40, 41, 43, 44, 45, 46, 47, 48, 49, 50, 51, 53, 54, 55, 57, 62, 65, 66, 67, 80, 81, 86, 99, 109, 110, 111, 113, 115, 126, 127, 129, 136, 137
Charles Darwin 21, 104, 107
Chemical particles ... 2, 136, 137, 138, 146, 149, 150, 161
Chemistry 137, 138
Children . 2, 93, 94, 98, 118, 143, 144, 145, 150
China 86, 90
Chinese 90
Christianity . 86, 87, 94, 106, 115, 128
Christians..2, 3, 86, 87, 89, 90, 92, 93, 95, 97, 122, 127, 141
Chronology 57, 87, 92, 161
Circumference 11, 62, 78
Confucianism 90
Confucius 90
Confusion 7, 51, 117
Cosmology 2, 7, 15, 16, 17, 18, 19, 38, 53, 84, 97, 102, 105, 108, 109, 111, 112, 114, 117, 119, 136, 140, 142, 147, 148, 162
Creation..2, 3, 4, 8, 17, 21, 84, 86, 87, 89, 90, 91, 92, 93, 94, 95, 96, 97, 98, 99, 100, 101, 102, 105, 106, 108, 111, 117, 120, 121, 126, 127, 131, 137, 139, 141, 142, 143, 146, 148, 149, 150, 151, 161
Creation stories 91
Creationism . 86, 92, 98, 99, 102, 122, 128, 129
Creationists 3, 7, 8, 12, 84, 85, 90, 91, 96, 97, 98, 99, 100, 101, 112, 115, 117, 120, 122, 128, 142
Creator .. 86, 105, 106, 112, 115, 120, 125, 127, 128, 129, 142, 146, 147
Crop Science Society of America 6, 161

D

Dale Carnegie .. 14, 15, 20, 83, 84, 96, 100, 103, 108, 116, 117
Daughter bodies 23, 25, 37, 38, 54, 64, 71
David J. Schwartz 20
David Yonggi Cho 115
Days of creation... 2, 87, 92, 102, 129, 140, 161
Death 111, 112
Demon 92
Discovery 145, 150
Distance11, 12, 13, 14, 15, 17, 19, 21, 24, 25, 26, 28, 29, 30, 31, 32, 33, 34, 38, 39, 40, 41, 42, 43, 44, 45, 54, 55, 57, 59, 62, 68, 69, 71, 74, 75, 80, 109, 110, 113, 126, 129
Doctoral Marshal 7
Donald Trump 107
Doubters 12, 104
Dr. Seuss 25
Drake 27

E

Earth ...2, 3, 11, 12, 13, 14, 15, 16, 17, 22, 23, 25, 26, 28, 30, 31, 32, 34, 35, 36, 37, 38, 39, 40, 41, 42, 43, 44, 45, 46, 47, 48, 50, 51, 52, 53, 54, 56, 57, 58, 59, 60, 63, 64, 66, 69, 70, 71, 72, 74, 75, 76, 79, 80, 81, 85, 87, 88, 92, 95, 98, 99, 100, 101,102, 109, 110, 111, 113, 115, 120, 121, 122, 126, 127, 129, 133, 139, 140, 142, 145, 148, 161
Earth's layers 34
Earth-Moon system22, 25, 26, 36, 37, 39, 42, 57, 59, 66, 69, 71, 72
Ecological Society of America ... 6, 161
Education. 6, 9, 12, 99, 100, 106, 112, 115, 117, 142, 143, 144, 147
Egypt 13, 94, 95
Elizabeth Stone 23
Embryo 24

Engineering 16, 113
Equations ... 29, 31, 38, 39, 40, 41, 42, 43, 45, 46, 47, 48, 50, 51, 52, 53, 54, 56, 57, 58, 59, 60, 62, 103, 105, 120, 126, 127, 142
Eratosthenes 13, 113
Escape velocity. 11, 15, 16, 17, 37, 38, 43, 64, 68, 69, 71, 74, 109, 110, 111, 126, 129
Evolution 2, 3, 7, 21, 85, 96, 98, 99, 100, 102, 104, 105, 106, 115, 117, 121, 122, 128, 138, 142, 144, 146, 148
Ex-believers 104
Experts 4, 5, 133, 138

F

Facts 3, 9, 16, 17, 18, 86, 103, 104, 109, 111, 112, 135, 143
Faith 2, 3, 6, 12, 56, 90, 93, 97, 99, 101, 102, 103, 106, 117, 121, 123, 127, 131, 133, 134, 140, 141, 142, 144, 145, 147, 148, 162
Fire 147
Firmament 88
Fluid layers 57
Fluid layers of Mercury's precursor 36
Fluid layers of the precursor of Mercury 66
Fluids 4, 22, 24, 28, 29, 33, 34, 35, 36, 37, 39, 40, 41, 42, 43, 44, 45, 46, 47, 48, 49, 50, 55, 57, 58, 59, 60, 62, 64, 65, 66, 67, 68, 69, 70, 71, 72, 74, 75, 77, 78, 80, 110
Forces 14, 86, 137
Forefathers 21, 92, 94, 113
Forerunners 21, 22, 26, 29, 38, 94
Forms of life 138
Fossils 35, 100
Framework (methodological) .. 4, 5, 7, 9, 11, 20, 25, 54, 56, 71, 84, 85, 96, 99, 101, 102, 106, 108, 109, 115, 116, 148

François Rabelais 113
Freethinkers 2, 12, 84, 85, 97, 98, 100, 101, 102, 103, 104, 115, 122, 136, 141, 142, 146, 147, 148
From Science to Bible's Conclusions 2, 87, 121, 139, 140, 145, 146, 149, 150, 151, 161

G

Galaxies 23, 42, 70, 80, 119, 137
Galileo Galilei 10, 26, 84, 86
Genesis ... 87, 90, 91, 96, 99, 101, 111, 126, 127, 129, 141, 142, 148
Gentiles 90, 93
Georg C. Lichtenberg 161
George W. Bush 107
George Washington 98, 132
God 2, 3, 4, 8, 9, 86, 87, 90, 91, 92, 93, 95, 96, 97, 98, 99, 100, 101, 102, 103, 104, 105, 106, 108, 111, 119, 120, 121, 122, 126, 127, 128, 129, 131, 134, 135, 136, 140, 141, 142, 143, 144, 145, 146, 147, 148, 149, 150, 151, 161
God created 87
God's Existence Scientific Formula 127
Gravity 11, 12, 14, 15, 16, 19, 37, 111, 137
Greeks 13, 14, 113
Greg Anderson 30
Gregor Mendel 115

H

Heaven 8, 87, 88, 99, 126
Hebrew 93, 112
Henri Poincaré 1, 104
Heraclitus 105
Hinduism 90
Hindus 91
How Baby Universe Was Born 2, 144, 145, 146, 149, 150, 151, 161
How God Created Baby Universe. 144

Human beings ... 8, 21, 22, 23, 24, 28, 95, 97, 102, 112, 113, 138
Humanists 12, 103

I

Ignorance 99, 106, 107, 109, 126, 128
In the beginning 64, 87, 126, 137, 142, 145, 161
Initial matter..... 22, 23, 24, 25, 54, 64
Intelligent Design 12, 85, 100, 101, 115
Intersection of science and faith 2, 106, 115, 121, 123, 127, 133, 134, 144, 147, 148, 149, 150, 161
Isaac Newton... 11, 12, 14, 16, 22, 32, 37, 51, 56, 68, 86, 91, 109, 110, 111, 117
Islam 3, 87, 91, 92, 93, 95, 96, 115, 128, 151
Israel.....2, 4, 3, 18, 28, 29, 30, 31, 32, 33, 37, 38, 39, 40, 41, 42, 43, 44, 45, 46, 47, 48, 49, 50, 51, 52, 53, 54, 57, 58, 59, 60, 61, 62, 64, 65, 66, 68, 69, 70, 71, 73, 74, 75, 77, 80, 81, 86, 90, 92, 93, 95, 96, 97, 113, 115, 120, 125, 126, 127, 129, 131, 134, 135, 139, 142, 148, 151
Israelites 93

J

James Allen 125
Jesus .87, 90, 91, 93, 95, 96, 112, 113, 114, 127
Jesus-Christ...... 87, 90, 91, 93, 95, 96, 112, 113, 114, 127, 128
Jewish 92, 95, 127
Jews ..3, 86, 87, 89, 90, 91, 92, 93, 95, 106, 112, 117, 120, 122, 127
Joe Biden 107
John Burdon Sanderson Haldane . 100
John F. Kennedy 107
John MacArthur 125
John Maxwell 96, 98, 107, 132
John the Apostle 126
Joseph Fourier 13
Journey 7, 9, 17, 21, 25, 27, 28, 29, 30, 31, 32, 37, 38, 42, 43, 44, 50, 51, 53, 54, 57, 59, 60, 65, 67, 68, 70, 74, 75, 109, 135, 136, 143
Judeo-Christian 91, 117, 120, 127
Jupiter 68, 70, 110, 113

K

Kepler 12, 14, 86
Knowledge 109, 115
Koran 87, 91, 92, 96, 115

L

Laboratory 5
Laypeople 7, 23, 32, 57, 118, 133, 139, 147, 161
Leonhard Euler 109, 113
Life 3, 6, 7, 9, 12, 24, 26, 28, 51, 67, 90, 91, 97, 99, 105, 106, 107, 109, 114, 115, 117, 118, 119, 120, 121, 122, 128, 133, 134, 135, 136, 138, 139, 140, 141, 142, 143, 144, 145, 146, 147, 148, 149
Loa Tsu 27
Lord Kelvin (William Thomson) 83, 115
Louis Pasteur 125

M

Mahatma Gandhi 12
Mark Twain 83
Mars 66, 68
Martin Luther King Jr. 30, 125
Mathematical Proof of God's Existence 2, 106, 121, 147, 148, 149, 150, 151, 161
Mathematics 2, 3, 8, 9, 10, 13, 21, 23, 26, 31, 32, 33, 40, 44, 47, 48, 53, 55, 56, 57, 60, 62, 69, 80, 86, 91, 93, 96, 102, 109, 111, 115, 120, 127, 133, 137, 140, 142, 148, 150, 161

Matthew Fontaine Maury...... 92, 100
Max De Pree................................ 100
Max Planck.................................. 116
Maya Angelou 4
Mercury 14, 15, 17, 35, 36, 38, 40, 43, 62, 66, 68, 78
Messiah............................91, 93, 127
Messianic 90, 93, 95, 115, 127, 128
Michael Faraday............................ 86
Milky Way Galaxy.......................... 42
Minerals 137
Mistakes. 84, 93, 97, 98, 99, 103, 108, 111, 131, 136, 143
Mohammed 87, 92, 93, 94
Molecules................................... 137
Moon . 2, 3, 11, 13, 14, 15, 16, 17, 22, 23, 25, 26, 28, 30, 31, 33, 36, 37, 38, 39, 40, 41, 42, 43, 44, 45, 46, 47, 48, 50, 51, 52, 53, 54, 56, 57, 59, 60, 63, 64, 66, 69, 70, 71, 72, 74, 75, 76, 79, 80, 81, 87, 92, 98, 109, 110, 113, 120, 121, 126, 127, 129, 140, 142, 145, 148, 161
Moses 87, 90, 91, 93, 94, 97, 127, 129
Motion 4, 12, 14, 37, 47, 48
Muslims..............3, 86, 87, 92, 93, 95
Myths 106

N

Napoleon at St Helena 98
NASA ... 4, 11, 16, 17, 18, 39, 131, 151
Nathanael-Israel Israel.....2, 4, 18, 28, 29, 30, 31, 32, 33, 37, 38, 39, 40, 41, 42, 43, 44, 45, 46, 47, 48, 49, 50, 51, 52, 53, 54, 55, 57, 58, 59, 60, 61, 62, 64, 65, 66, 68, 69, 70, 71, 72, 73, 74, 75, 76, 77, 78, 80, 81, 86, 90, 93, 96, 113, 114, 115, 120, 125, 126, 127, 129, 131, 133, 134, 135, 137, 140, 141, 142, 144, 145, 146, 147, 148, 149, 150, 151, 161, 162
Nathanael-Israel Israel Birthdate .. 30, 31, 51, 52, 53, 54, 55, 57, 60, 61, 62, 66, 78, 80, 81, 86, 115, 126, 129, 131
Nathanael-Israel Israel Birthdate of the Earth 58
Nathanael-Israel Israel Birthdate of the Moon 60, 76
Nathanael-Israel Israel Birthdate of the Sun.......................... 61, 62, 78
Nathanael-Israel Israel Gathering Distance. 44, 45, 46, 47, 49, 50, 64, 72, 75, 78, 131
Nathanael-Israel Israel Gathering Speed45, 47, 48, 50, 64, 72, 75, 78, 126, 129, 131
Nathanael-Israel Israel Gathering Time. 30, 31, 32, 43, 44, 45, 46, 49, 50, 51, 52, 53, 55, 57, 60, 62, 66, 72, 75, 78, 131
Nathanael-Israel Israel Linear Distance. 32, 33, 37, 39, 40, 44, 68, 69, 71, 74, 126, 129, 131
Nathanael-Israel Israel Linear Journey 28, 29, 30, 32, 37, 65
Nathanael-Israel Israel Linear Speed ... 32, 33, 37, 38, 40, 44, 64, 69, 71, 74, 126, 129, 131
Nathanael-Israel Israel Linear Time ... 30, 31, 32, 33, 37, 39, 40, 41, 42, 43, 44, 46, 51, 52, 53, 55, 57, 59, 62, 65, 68, 69, 70, 72, 74, 75, 77, 131
Nathanael-Israel Israel Turbulent Journey 29, 65
Neptune 66, 68, 70
New Testament................. 93, 94, 95
Niels Bohr...................................... 84
Nobel Prize............................ 17, 148
Nonbelievers 8, 12, 103
Non-Messianic 93, 127
Nonreligious.................. 12, 103, 120

Nonscientists..... 7, 67, 108, 113, 122, 128, 138, 141
Nontheists 103

O

Old Earth creationism......... 12, 98, 99
Old Testament 93
Oprah Winfrey 105
Orbital speed ... 11, 12, 13, 17, 19, 43, 47, 48, 50, 57, 64, 72, 75, 78, 109, 126, 129
Origin of chemical particles..... 2, 136, 137, 146, 149, 150, 151, 161
Origin of life..... 2, 138, 139, 146, 149, 150, 151, 161
Origin of Life 151
Origin of the Spiritual World 146

P

Paul the Apostle 95, 116
Pharisees 94
Philosophy 90, 118, 136, 137
Physics 5, 15, 16, 17, 37, 39, 45, 94, 105, 109, 110, 111, 127, 136, 148
Planets..12, 15, 23, 24, 25, 37, 48, 67, 70, 80, 110, 113, 119, 136, 137, 151
Pluto 66, 68
Politics 147
Precursor of a primary body..... 26, 37
Precursor of a system of celestial bodies 26
Precursor of all the bodies orbiting the Sun 43, 64, 110
Precursor of Mercury ... 35, 36, 43, 68
Precursor of the bodies orbiting the Sun 66
Precursor of the Earth . 22, 36, 37, 38, 39, 42, 57, 58, 67, 70, 71, 72, 74, 110
Precursor of the Earth-Moon System 38, 53, 67, 68, 70, 71, 110
Precursor of the Moon 37, 38, 39, 42, 53, 59, 60, 70, 71, 74, 75, 76, 80, 110
Precursor of the Solar System 64
Precursor of the Sun 36, 37, 38, 39, 42, 43, 62, 64, 65, 67, 68, 69, 75, 78, 110, 113
Precursor to the Earth-Moon System 36, 67, 69
Precursors of bodies 21, 22, 23, 24, 25, 26, 27, 28, 29, 30, 31, 32, 33, 34, 35, 36, 37, 38, 39, 40, 41, 42, 43, 44, 45, 46, 47, 48, 49, 50, 53, 54, 55, 57, 58, 59, 60, 62, 64, 65, 66, 67, 68, 69, 70, 71, 72, 74, 75, 76, 77, 78, 80, 109, 110, 111, 113, 137
Predecessors 21
Pregnancy 23, 24
Primary bodies. 11, 25, 26, 36, 37, 39, 40, 41, 55, 126, 129
Proof.....7, 8, 9, 12, 48, 85, 90, 91, 93, 97, 99, 101, 102, 103, 104, 108, 118, 119, 127, 131, 133, 134, 147, 148, 161
Prophets 93
Pythagoras 10

Q

Questions ...4, 2, 3, 4, 7, 9, 16, 17, 18, 19, 22, 25, 38, 85, 86, 104, 114, 115, 118, 127, 131, 133, 134, 136, 137, 141, 143, 144, 146, 147, 148
Quran 92

R

Radius .. 11, 13, 14, 17, 19, 46, 57, 59, 61, 64, 72, 75, 78, 109, 113, 126, 129
Rationalists 2, 3, 8, 12, 14, 17, 90, 96, 101, 102, 103, 104, 106, 108, 111, 127, 134, 139, 140, 141, 142, 144, 146, 147, 148

Reality 16, 90, 95, 100, 108, 109, 121, 144
Reason 3, 5, 8, 13, 17, 27, 38, 84, 101, 102, 110, 121, 122, 125, 127, 128, 134, 137, 140, 141, 144, 147, 148, 162
Reconciling Science and Creation Accurately 2, 8, 87, 91, 95, 98, 102, 121, 128, 141, 146, 149, 150, 151, 161
Reference..........25, 39, 53, 59, 60, 74
Religion 2, 3, 5, 6, 8, 16, 17, 21, 29, 32, 84, 85, 86, 87, 90, 91, 93, 99, 101, 102, 103, 104, 106, 108, 111, 112, 114, 115, 117, 119, 120, 121, 123, 127, 128, 129, 133, 140, 142, 144, 147, 148, 162
Religiously unaffiliated......12, 86, 104
Research 3, 4, 9, 18, 30, 54, 80, 90, 103, 115, 118, 119, 121, 122
Revelation 3, 92, 93, 106, 115, 127, 128, 129, 131, 142, 145, 146, 161
Richard P. Feynman 105, 117
Ridiculous.................................... 111
Roadside showing soil layers 34
Rocks... 137

S

Sam Ewing.................................... 116
Satellites .. 15, 23, 25, 48, 70, 80, 119, 136, 137, 151
Saturn 66, 68, 70, 110, 113
School board of education........... 106
Science ... 2, 3, 4, 6, 7, 8, 9, 12, 32, 56, 90, 93, 95, 96, 97, 99, 100, 101, 102, 103, 106, 107, 108, 109, 114, 115, 117, 120, 121, 122, 123, 127, 128, 131, 133, 134, 136, 138, 139, 140, 141, 142, 144, 145, 147, 148, 150, 161
Science1802, 3, 4, 133, 136, 138, 139, 140, 142, 143, 145, 146, 147, 149, 150, 151, 161, 162
Science180 Academy ...12, 38, 62, 91, 128, 133, 134, 136, 141, 161
Science180 Accurate Scientific Proof of God 90, 92, 93, 95, 98, 102, 128, 147, 161
Science180 Consulting 134
Science180 Creationism...... 101, 102, 122, 142
Science180 Interview.................. 134
Science180 Masterclass 134
Science180 Publishing............. 4, 134
Science180 Seminars 134
Science180 Turbulence 1 .... 105, 109, 119, 121
Scientific Biblical Inerrancy Coefficients............. 126, 127, 129
Scientific data... 3, 4, 5, 6, 7, 9, 11, 17, 18, 19, 23, 26, 34, 35, 37, 39, 43, 48, 52, 54, 57, 61, 62, 80, 85, 97, 100, 101, 105, 106, 107, 113, 114, 118, 119, 121, 125, 126, 131, 133, 136, 138, 139, 140, 141, 142, 145, 148, 161, 162
Scientific evidence .10, 17, 80, 87, 92, 102, 104, 115, 128, 145
Scientific formula 4, 108, 113, 126, 129, 135, 136, 138, 140, 141, 149, 150
Scientific theory ...136, 137, 139, 140, 141, 142, 143, 144, 147, 161, 162
Scientifically verifiable 3, 101, 148, 161
Scientists 5, 6, 7, 11, 14, 15, 17, 22, 23, 26, 31, 32, 37, 45, 54, 62, 68, 103, 105, 106, 107, 108, 109, 110, 111, 112, 113, 114, 118, 119, 121, 122, 125, 126, 128, 133, 136, 139, 140, 141, 142, 148, 161
Secondary bodies...25, 26, 36, 37, 39, 40, 55
Secular . 2, 84, 85, 100, 101, 105, 106, 107, 108, 112, 117, 120, 121, 122, 142, 144

Semi-major axis...... 11, 14, 15, 19, 39
Six 24-hour days...................... 87, 92
Six-day 24-hour 3
Skeptics . 12, 103, 136, 139, 140, 141, 142
Soil Science Society of America.... 161
Solar System.... 11, 12, 18, 22, 24, 25, 26, 30, 37, 38, 39, 42, 51, 53, 59, 60, 64, 66, 68, 70, 72, 74, 76, 78, 80, 82, 110, 151
Souls 111
Space 7, 11, 13, 14, 15, 17, 22, 23, 31, 37, 106, 110, 111, 131
Spaghetti noodle 46, 47, 48, 49, 50
Speaking Engagement 134
Speed....11, 13, 15, 16, 19, 21, 22, 24, 25, 26, 28, 29, 30, 31, 32, 33, 34, 36, 37, 38, 39, 41, 44, 45, 46, 47, 48, 49, 50, 54, 55, 62, 67, 70, 77, 78, 80, 110, 113
Spirit 115, 127
Spirit of God 88
Spiritual 2
Spiritual world ... 2, 95, 145, 149, 151, 161
Spiritual-not-religious............ 12, 104
Split-gathering. 22, 23, 24, 25, 26, 28, 29, 32, 34, 36, 37, 39, 42, 43, 44, 45, 46, 47, 49, 54, 64, 65, 66, 67, 68, 74, 78, 88, 111
Stars.13, 23, 80, 88, 92, 119, 135, 137
Steve Jobs 2
Sun 2, 3, 11, 12, 13, 14, 15, 16, 17, 22, 23, 25, 26, 28, 30, 31, 33, 35, 36, 37, 38, 39, 40, 41, 42, 43, 44, 45, 46, 47, 48, 50, 51, 52, 53, 54, 56, 57, 59, 60, 61, 62, 63, 64, 65, 66, 67, 68, 69, 70, 71, 74, 77, 80, 81, 87, 92, 98, 109, 110, 111, 113, 120, 121, 126, 127, 129, 141, 142, 145, 148, 151, 161
Sunday.................................... 87, 92
Sweden.. 151

T

The Word............................ 112, 126
Theories....2, 3, 4, 5, 7, 11, 12, 14, 15, 16, 17, 18, 21, 23, 32, 38, 80, 83, 84, 85, 87, 90, 92, 97, 98, 99, 101, 102, 103, 104, 105, 106, 108, 109, 110, 111, 113, 114, 115, 117, 118, 119, 120, 121, 122, 125, 127, 128, 129, 148
Think.2, 5, 6, 8, 12, 14, 15, 16, 17, 20, 31, 32, 48, 85, 86, 90, 91, 93, 95, 96, 98, 100, 101, 103, 104, 106, 108, 110, 111, 112, 113, 114, 115, 116, 117, 128, 136, 142
Thinkers 12, 14, 26, 115, 141
Thomas A. Edison 21
Time....2, 3, 4, 5, 6, 7, 8, 9, 12, 16, 17, 21, 23, 24, 28, 29, 30, 31, 32, 39, 42, 43, 44, 45, 46, 47, 49, 50, 51, 53, 54, 57, 58, 59, 60, 62, 63, 65, 66, 68, 70, 73, 74, 75, 77, 81, 85, 86, 90, 94, 96, 102, 103, 104, 105, 106, 107, 108, 109, 110, 111, 115, 117,118, 119, 131, 136, 139, 140, 142, 143, 149, 161
Tony Robbins 27
Top questions 133
Tradition 90, 91, 120, 148
Turbulence 2, 4, 5, 6, 7, 17, 19, 21, 24, 25, 28, 37, 44, 45, 47, 48, 54, 55, 64, 65, 70, 71, 81, 109, 114, 119, 121, 126, 127, 135, 136, 137, 138, 139, 146, 148, 149, 150, 151, 161
Turbulent Origin of Chemical Particles........ 24, 64, 121, 149, 150
Turbulent Origin of Life 121, 139, 149, 150
Turbulent Origin of the Universe. 2, 8, 11, 16, 22, 24, 26, 27, 31, 33, 35, 38, 40, 42, 45, 51, 55, 64, 69, 70, 80, 111, 114, 121, 135, 136, 146, 151, 161
Turbulent Prima Materia 22, 24, 64

U

Unbelievers.... 2, 8, 85, 94, 95, 96, 98, 99, 101, 102, 103, 104, 108, 121, 147
Understanding . 5, 6, 7, 11, 28, 31, 32, 37, 48, 51, 56, 101, 106, 111, 115, 117, 118, 120, 122, 131, 134, 136, 137, 138, 139, 140, 141, 144, 145, 146, 147
Universe . 2, 3, 4, 5, 7, 8, 9, 10, 11, 12, 14, 15, 16, 17, 18, 21, 22, 23, 24, 25, 27, 29, 31, 32, 35, 44, 45, 51, 54, 64, 70, 71, 80, 84, 87, 90, 92, 94, 95, 96, 97, 99, 101, 104, 105, 106, 108, 109, 110, 112, 113, 114, 115, 117, 118, 119, 120, 121, 127, 128, 129,131, 134, 135, 136, 137, 138, 139, 140, 141, 142, 143, 144, 145, 146, 147, 148, 161
Uranus..................................... 66, 70
Ursula K. Le Guin 29
USA.............4, 6, 7, 13, 128, 151, 162

V

Venus 36
Virgin Mary 93
Vortices 4, 45

W

Waters... 88
Werner Heisenberg...................... 103
William K. George 5, 16, 103
William Paul Thurston.................... 56
Winston Churchill........................... 83
Wisdom.....................83, 84, 114, 115

Y

Yeshua HaMashiach......... 90, 93, 128
Young Earth creationism... 85, 98, 99, 100, 101

About the Author

"It is strange that only extraordinary men make the discoveries, which later appear so easy and simple." — Georg C. Lichtenberg

Dr. Nathanael-Israel Israel is privileged to be the discoverer of the simple, all-in-one, scientifically verifiable formula that mathematically demonstrated that the Earth was formed 2.82 days, the Moon 3.32 days, and the Sun 3.69 days after the beginning of the universe, therefore confirming the Biblical account of creation that revealed about 3500 years ago that the formation of the Earth was completed on the 3rd day, while that of the Moon and the Sun was completed on the 4th day of creation. He is the author of *From Science to Bible's Conclusions*, *Turbulent Origin of the Universe*, *Reconciling Science and Creation Accurately*, *Turbulent Origin of Chemical Particles*, *Turbulent Origin of Life*, *How Baby Universe Was Born*, *Origin of the Spiritual World*, *How God Created Baby Universe*, *Science180 Accurate Scientific Proof of God*, and *Mathematical Proof of God's Existence at the Intersection of Science and Faith'.* Founder of Science180 (www.Science180.com) and of Science180 Academy (www.ScienceAcademy.com), he is a member of the American Association for the Advancement of Science, the American Chemical Society, the American Society for Microbiology, the American Society of Biochemistry and Molecular Biology, the Ecological Society of America, the American Society of Agronomy, the American Society of Agricultural and Biological Engineers, the Crop Science Society of America, the Soil Science Society of America, the American Physical Society, the American Mathematical Society, the American Philosophical Association, and many others, including renowned religious, theological, associations / societies. See the full list at Israel120.com/affiliations. These memberships also underscore that his work is grounded in rigorous scientific standards. He recognizes that, in the modern world, science is often used to argue against God, and he is committed to demonstrating that actual science need not deny God's existence.

Since its invention, science has significantly benefited humankind, yet it has also become a modern means by which some people deny God's existence. Many scientists, non-scientists, and laypeople from various backgrounds have advanced numerous theories claiming to prove or disprove God's existence. However, based on the evidence presented, it is clear that no one has yet examined the origin of the universe and its contents through the unconventional scientific perspective that Dr. Nathanael-Israel Israel has discovered. The mathematical proof decoded by Dr. Israel not only proposes a new theory but also uses existing peer-reviewed scientific data to, for the first time in history, arrive at an astounding chronological confirmation of the Biblical creation account. It provides a precise scientific timeline that resolves centuries of debate, offering clarity at the fundamental intersection of cosmology and scripture. Dr. Israel uncovered overlooked datasets containing answers long sought by both scientific and theological communities. Dr. Nathanael-Israel Israel, who completed his BS and MS in Africa—finishing first in his class—before moving to the USA, where he earned a PhD, again finishing first among hundreds of peers. Recognized as a distinguished scientist, he is also a believer. He does not oppose science; instead, he opposes certain scientific and religious methods that hinder the pursuit of truth. While he does not believe science is against God, he acknowledges that many scientific and religious theories conflict with Him, even when based on substantial data. He is therefore pleased to have found an intersection of science and faith where reason and faith can coexist without denying God or rejecting raw scientific data, including that which underpins controversial theories.

To learn more about this rare scientist and believer, please visit his personal website, www.Israel120.com, and the media kit of Science180, www.Science180.com/mediakit